The EFFECTS of ULTRASOUND on the KINETICS of CRYSTALLIZATION

A. P. Kapustin

Authorized translation from the Russian

CONSULTANTS BUREAU
NEW YORK
1963

ISBN-13: 978-1-4684-1550-6   e-ISBN-13: 978-1-4684-1548-3
DOI: 10.1007/978-1-4684-1548-3

The Russian text was published by the USSR
Academy of Sciences Press for the
Institute of Crystallography, in Moscow, in 1962.

Editor-in-Chief
Academician A. V. Shubnikov

Effect of Ultrasound on the Kinetics of Crystallization

VLIYANIE UL'TRAZVUKA NA KINETIKU KRISTALLIZATSII

Александр Павлович Капустин

Влияние ультразвука на кинетику кристаллизации

Library of Congress Catalog Card Number 63—17640

Contents

Contents

Introduction

One of the important new research methods involves the use of ultrasonics during crystallization or dissolution.

Ultrasound has become extensively used in many branches of science and technology in recent years, as we may see from the number of papers published. The total up to 1939 was about 700, but by 1960 it was well over 5000. Applications in physics, crystallography, chemistry, and so on provide a powerful means of discovering new effects, although such studies tend to be very complicated and demand an acquaintance with a wide range of topics. The main effect of ultrasound on matter is to change the energy state considerably; an ultrasound flux as low as 10 W/cm^2 at 3×10^6 c/s produces a pressure wave whose amplitude is 5 atm, while the maximum acceleration may exceed 10^5 times that of gravity, the maximum particle speed being 40 m/sec and the radiation pressure 1300 dynes/cm^2 [1, 2].

Recent ultrasonic emitters, even without focusing systems, can give steady fluxes up to 300 W/cm^2.

The kinematic and dynamic parameters of ultrasonic fileds are such that they can cause effects at the level of atoms and molecules, quite apart from macroscopic effects. Cavitation is the most important of the various effects caused by an intense ultrasonic field; the main features of this are firstly that cavities are produced in the liquid, and these are filled by the vapor and by the gas dissolved in the liquid, and secondly that intense shock waves are generated when these cavities collapse [3]. The pressures in these shock waves are much in excess of those caused by the ultrasonic waves directly; they are sufficiently high to disrupt solids. The effects produced by ultrasound also include 1) dispersion of solids and liquids; 2) coagulation and precipitation of suspensions; 3) mass transfer in liquids and gases; 4) chemical reactions; and 5) alterations in processes of crystallization and dissolution.

Ultrasound has become widely used in crystallographic work in the last 15 years; it has been applied to monocrystals and to crystal assemblies, as well as in studies of crystallization and of crystal growth and dissolution.

Crystal growth is an important technical problem in many fields, on account of the steadily increasing uses of crystals; studies on nucleation, growth, and dissolution, in particular as regards the effects of various agents on the kinetics of the phase transition, are thus of considerable importance. One such agent, as S. Ya. Sokolov has shown, is ultrasound, although its effects on crystallization and dissolution are complex. The variety of factors that can affect nucleation, growth, and structure (even in the absence of ultrasound) is such that it is difficult to construct a unified theory capable of explaining all effects [4]. All the same, many useful results have been obtained from the use of ultrasound, which has been found to alter the primary crystallization structure; this is of great importance to the production of materials with improved physical or chemical properties. Ultrasound can alter the grain size, can produce a zonal structure, can alter or minimize the zone of columnar crystals, and so on.

Further, ultrasound is of value in examining existing structures; the attenuation coefficient is, for example, very much dependent on the grain size, on the texture (if any), and so on.

It for long remained an open question whether ultrasound affects the growth of a monocrystal, but recent work has shown that it may accelerate, retard, or not affect the growth rate (the supersaturation, frequency, and intensity are the decisive factors here). Moreover, these effects are also dependent on the size of the seed and on the position of the face relative to the wave front. This provides a means of controlling rates of crystallization and dissolution.

Experiments have also been made on the effects of ultrasound on the growth of metal monocrystals; for example, a zinc crystal grown in an ultrasonic field can have improved mechanical characteristics (the elastic limit may be raised by a factor of six).

New results have been reported for nucleation; here the effects of ultrasound are dependent on the purity and also on any previous ultrasound treatment given while the material was in a superheated or supercooled state.

A very recent application is to the production of etch figures.

Some effects have already found practical uses although they have not been fully elucidated. Clearly, a careful practical and theoretical study of effects on crystallization is needed in order to provide the basis for future fruitful uses.

One of the main tasks in research on phase transitions is to find means of influencing crystallization processes in order to give the final crystals the desired properties.

Industrial requirements demand experimental and theoretical researches on this complex problem, which is closely related to that of controlling phase transformations. Research on new methods of influencing crystallization (dissolution) should provide the means of making materials with properties specified in advance.

This book deals mainly with the joint work of myself and my colleagues on the experimental side of crystallization and dissolution in ultrasonic fields of various frequencies and strengths; some effects produced in crystalline materials are also described.

There is also a brief survey of the principal experimental work on the kinetics of crystallization in ultrasonic fields.

The book is directed to those specializing in the nucleation, growth, and dissolution of crystals, and also to instructors in higher educational institutions; it is hoped that it may stimulate new lines of research.

I am much indebted to Academician A. V. Shubnikov for advice on many of the topics studied and also on the compilation of this book. I am also indebted to Professors G. G. Lemmlein and B. B. Kudryavtsev for valuable suggestions made upon reviewing the book in manuscript.

I should be grateful to be notified of any deficiencies anyone may encounter in this book.

CHAPTER I

Methods and Apparatus for Studying Crystallization and Dissolution in Ultrasonic Fields

1. General

Crystal growth is governed by the nature of the substance and by the crystallization conditions; many processes are involved in the initiation and growth of a crystalline phase in a liquid or vapor, but the initial stage (nucleation) is the decisive one in the formation of a solid from a melt or solution and in the recrystallization of a solid in response to physical factors.

There are two related but distinct effects in crystallization, namely nucleation of the new phase and subsequent growth. The linear growth rate and the number of nuclei produced in unit volume in unit time are the kinetic parameters that determine the final polycrystalline structure. The position of the maximum on the curve of nucleation rate as a function of supercooling determines whether the solid product is obtained in the crystalline or amorphous state.

Ultrasonic methods have provided new means of controlling the kinetics of phase transformations; the field strength can be adjusted to control the number of nuclei while leaving other factors unaltered, so the rate of crystallization can be controlled.

Many observations have been made on crystallization processes as influenced by external fields (electrical, magnetic, sonic, ultrasonic); radioactive emissions have also been used. The results may be summarized by saying that radiations and electric and magnetic fields all facilitate nucleation. Shubnikov [5] has recently demonstrated the effects of an electric field on nucleation in solutions of ammonium chloride. Yasuchi's [6] results are interesting; he found that electromagnetic waves of $\lambda = 30\text{-}90$ cm facilitate nucleation in solutions of salts.

There are many papers on the effects of vibrations (in particular, sound and ultrasonics) on crystallization; it is considered that cavitation is mainly responsible for the large increase in the number of nuclei. Kuznetsov [7] and Hiedemann [8] have reviewed vibration effects.

Considerable interest attaches to the interaction between ultrasound and a crystallizing or dissolving material, for many aspects of molecular physics are involved (nature of molecular forces, theory of thermal capacity, and so on). There are now many papers on these topics [9-11].

2. Choice of Material for Observing Crystallization

Nucleation can be observed under the microscope, though this restricts the work to transparent substances and excludes metals. Special requirements have to be met if particular stages are to be observed and photographed. The most suitable substances are transparent ones that have low melting points and that are readily supercooled; they should also not decompose on prolonged use and should have low linear rates of crystallization. Frequently used substances are as follows: thymol, piperonal, menthol, salol, benzophenone, o-chloronitrobenzene, benzyl, azobenzene, p-bromaniline, and nitrobenzene. Some of these (e.g., o-chloronitrobenzene) have pronounced limits of metastability, so their behavior is similar to that of metals, which do not give high degrees of supercooling.

The substance must be purified, for impurities can have a pronounced effect. Methods of purification include filtration, centrifugation, and repeated crystallization from solution. Danilov and others [12] have given good descriptions of impurity effects and of purification methods.

3. Ultrasonic Waveguides and Radiators

Electromechanical converters and special waveguides are needed to produce and transport ultrasonic energy; here I shall describe mainly the apparatus designed and tested in the laboratories of the Institute of Crystallography, Academy of Sciences, USSR. References to novel apparatus designed elsewhere are given in the text.

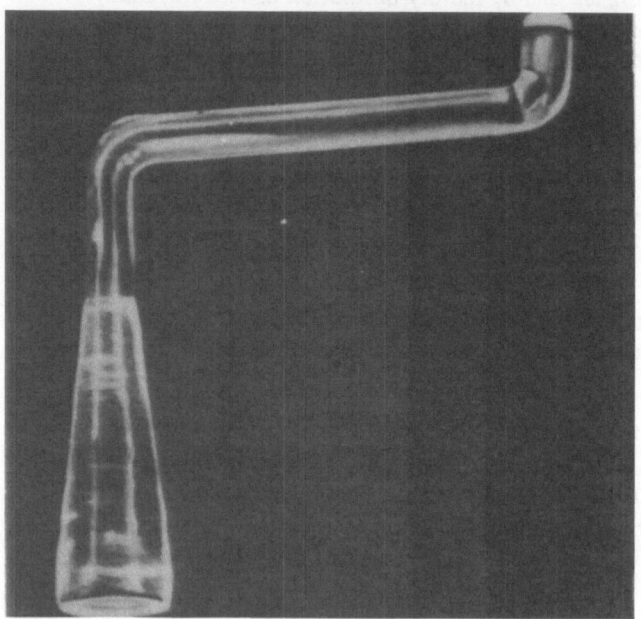

Fig. 1. Glass waveguide.

A simple glass waveguide in the form of a bent rod (Fig. 1) can be used with thin layers of supercooled liquids; the wide end of the rod is the input, and the narrow end forms the object stage, which is readily inserted under the microscope. The best position of the wide end relative to the source is found by experiment; the waveguide is moved vertically with a micrometer screw. However, this waveguide causes fairly large energy losses on account of reflection at the output end.

Experiments at controlled temperatures are performed in a thermostat containing the source. A brass vessel with double walls carrying a liquid at a fixed temperature is used; this is 500 mm high and 25-30 mm in outside diameter. Figure 2 shows the design of the source, which is placed at the bottom. A quartz plate is metallized on both sides, apart from a circle 3-4 mm in diameter in the center, which is left clear to pass a light beam through the specimen to the microscope. A glass cube placed directly on the quartz plate forms the object stage. Acoustic contact between plate and cube is improved by a few drops of oil; the inner brass vessel is also filled with oil sufficient to cover the quartz plate if high voltages are to be used. The sample on the cube is heated and cooled by adjusting the temperature of the jacket, the temperature of the liquid being kept constant to 0.1°C. The temperature at the specimen is measured with a thermocouple inserted in a hole near the surface of the glass cube. The system provides vertical and horizontal ultrasonic beams; in the latter case, the quartz plate is fixed to the side of the vessel.

A special holder has been designed for use at high powers; Fig. 3 shows this. The lead block 80 mm in diameter and 20 mm high has a recess 3-4 mm deep at the center; near the edge there is a rim bearing a bronze foil (0.05 mm thick) tensioned by a brass ring. On one side of the foil there is air, which acts as a good reflector for ultrasonics.

The quartz plate is placed on the foil, and on the plate there is placed a polished brass ring (external diameter 58 mm, internal 48 mm, thickness 2 mm). The high-frequency voltage is fed in through flexible brass plates, which press on the brass ring. The pressure and the orientation of the plates have pronounced effects on the amplitude of vibration. The best effect is obtained with curved plates pressing on the brass ring from two sides at four points. The pressure is chosen by experiment; it should be fairly high. The brass plates are held on glass rods by bosses and screws.

Various devices have been described for studies on crystallization in bulk phases. Figure 4 shows one such, in which a quartz vessel with double walls provides means of rapidly heating and cooling the material. Figure 5 shows Bagdasarov's device. A tube containing the substance is inserted into the thermostat via rubber sealing rings; the tube is 10 mm in diameter and has a base 0.5 mm thick. A film of perfol 0.05 mm thick is used as the base if the ultrasonic frequency is several megacycles.

8

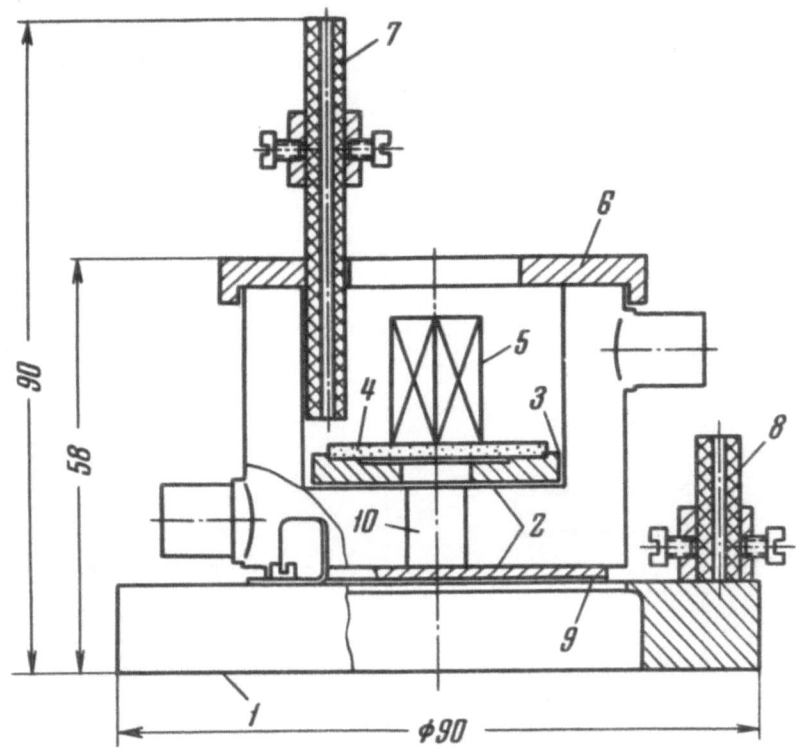

Fig. 2. Thermostat with quartz holder: 1) Base; 2) metal cylinder with
double walls; 3) lead plate; 4) quartz; 5) glass cube; 6) cover; 7) porcelain
tube; 8) current lead; 9) glass base; 10) hole to admit light beam.

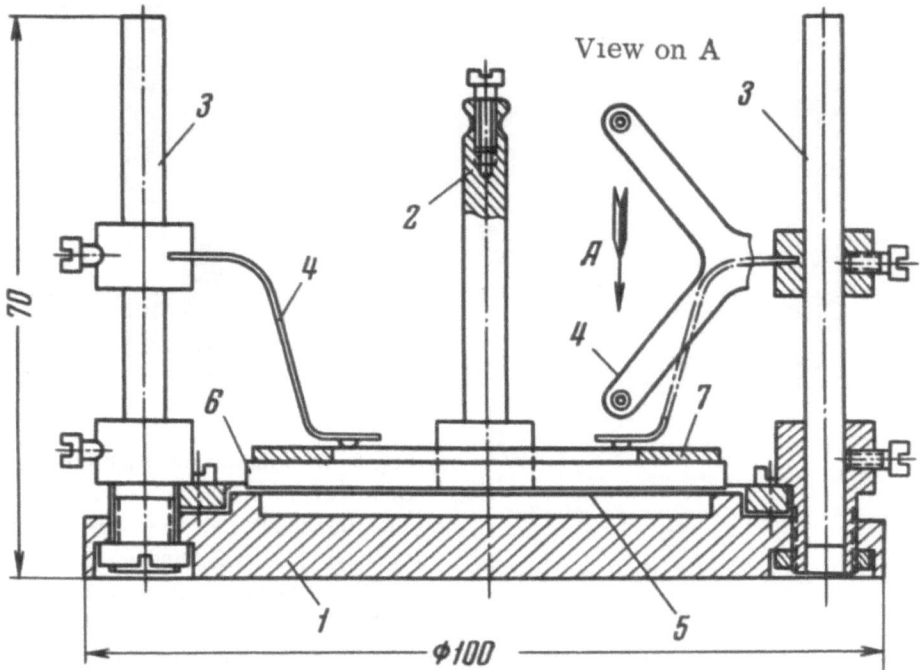

Fig. 3. Holder for quartz plate: 1) Base; 2) metal rod to carry current; 3) glass rod;
4) contact plate; 5) bronze foil; 6) quartz; 7) brass ring.

9

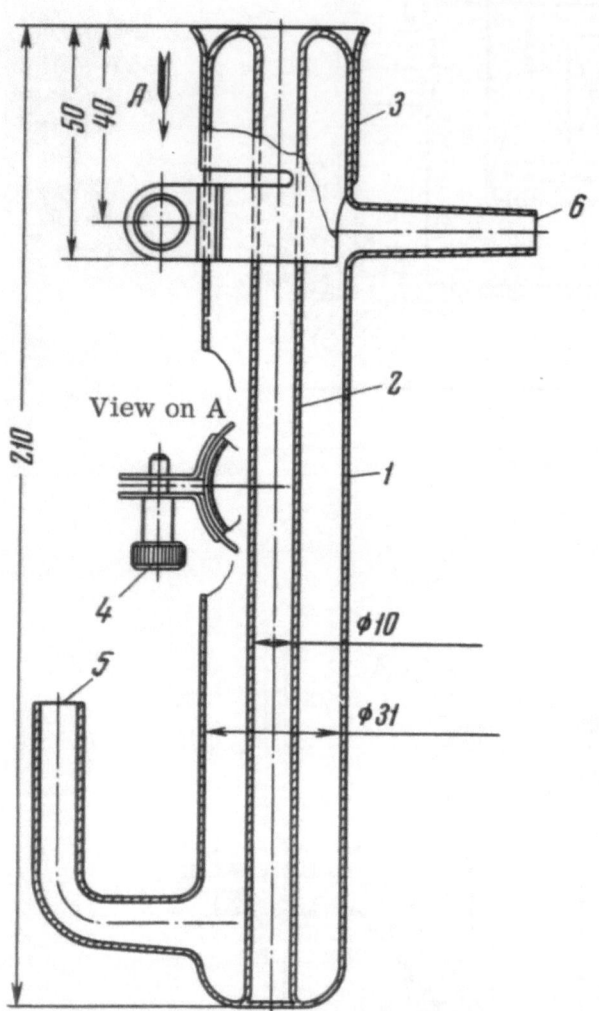

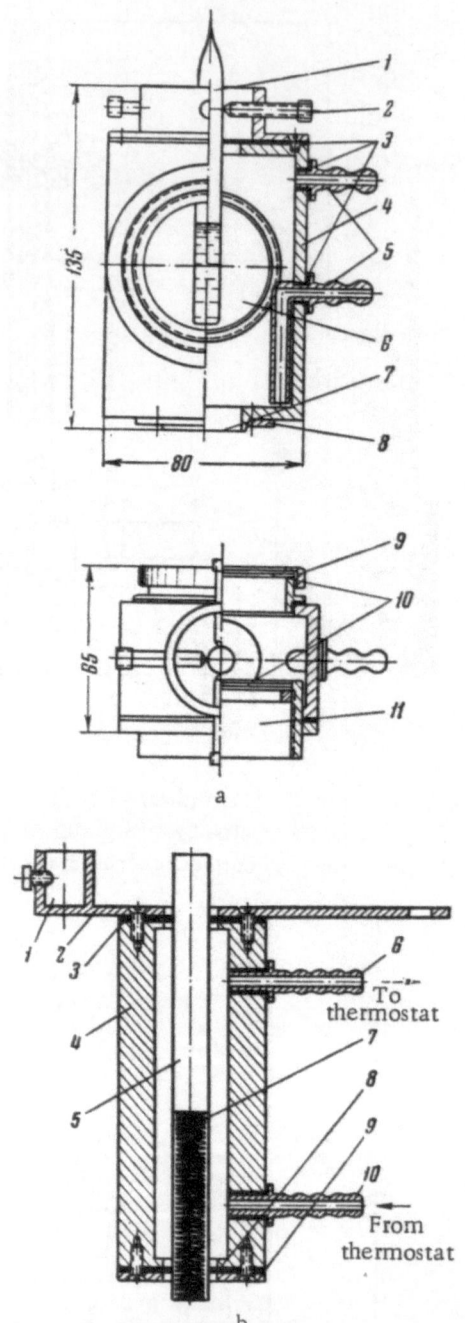

Fig 4.   Quartz tube for studying phase-transformation rates in the bulk: 1 and 2) Outer and inner walls of tube; 3) boss for clamping tube; 4) clamp; 5 and 6) holes for water.

Fig. 5. Thermostat for examining crystallization in an ultrasonic field. Part a): 1) Tube containing material; 2) screw; 3) rubber seals; 4) body; 5) tube; 6) window; 7) perfol; 8) ring; 9) fixed window; 10) glass; 11) mobile window. Part b): 1) Recess for probe thermocouple; 2) upper cover; 3 and 8) rubber seals; 4) body; 5) glass tube; 6 and 10) inlet and outlet tubes; 7) material; 9) lower cover.

A different type of system is needed for work with liquid metals. It is difficult to design waveguides for this application, because we lack data on the speed and absorption as functions of temperature for liquid metals. The design must provide high temperatures and maintenance of such temperatures; it must also protect the source from undue heating while ensuring the most efficient use of the radiated energy. The crucible must transmit the waves freely but must not form alloys with the metal. Various methods have been used; the topic is considered in detail by Bergmann [1] and Crawford [13].

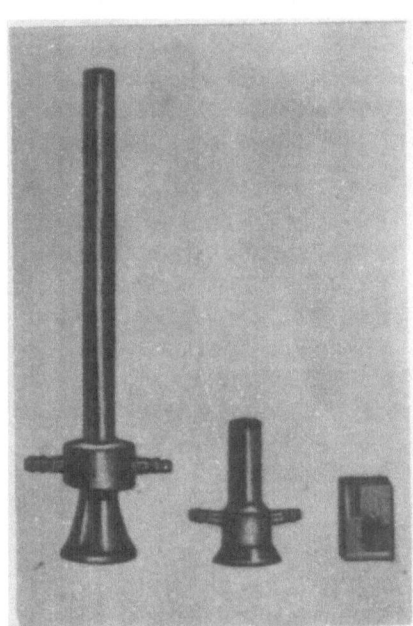

Fig. 6. Metal waveguides.

Figure 6 shows several waveguides that have been used as thermal insulators in conjunction with crystal oscillators. Each is a cylinder whose length is adjusted by trial, the lower end being a cone to match the diameter of the quartz plate (60 mm). The upper part has a recess in the form of a truncated cone, in which the metal is melted. The rod, if of metal, is fitted at the lower end with a water-cooled jacket. The metal sample is melted in a tube oven placed over the upper end of the waveguide. The broad end of the waveguide is placed above the quartz plate, which is immersed in oil. The position is adjusted to give maximum power transfer to the rod.

A conical concentrator between the source and the sample can provide a useful increase in power density in the metal. Teumin has used this method.

Recently, practical calculations on concentrators have been given by Merkulov [14] and by Nagol'nykh and Rozenberg [15].

CHAPTER II

A Brief Review of Work on the Interaction of Ultrasonic Energy
with Crystallizing or Dissolving Material

1. Crystallization of Various Materials in an Ultrasonic Field

Here I deal first with some work on the effects of ultrasonic energy on crystallization; next I deal with some results, most of which have been obtained at the Institute of Crystallography. The review is restricted to work directly relating to the studies presented here.

Some interesting and valuable effects occur when a melt, solution, or solid is exposed to ultrasound.

The grain size of a melt solidifying at a normal rate is governed by many factors, such as the rate of cooling, the nature of the material, and the presence of impurities [4]. It is common knowledge that the microstructure is important.

All the main physicochemical properties of disperse systems are concerned here, especially adsorption, catalysis, and chemical activity; these are widely used in practice.

The general effect of insonation is to produce a fine-grained structure, on account of extensive nucleation; but the growth of monocrystals is sometimes facilitated. High intensities cause cavitation, which facilitates depolymerization, oxidation, and removal of gases; the last is important in the improvement of the structure. Ultrasonic methods are of considerable value in the production of homogeneous optical glass, for example; gas removal by ultrasonic techniques is used with organic materials and molten metals.

Richards and Loomis [16] were the first to observe the effects of ultrasonic energy on crystallization; they found that a supersaturated solution of thiosulfate is caused to crystallize.

Loomis and Wood [17] described the instant crystallization of thiosulfate in response to ultrasonic energy transmitted through a glass rod.

Rapid crystallization also occurs if the solution is seeded with a crystal of thiosulfate and the ultrasonic beam is simultaneously turned on.

Danilov et al. [18] examined the effects on piperine at 1.8 Mc; the number of crystallization centers formed in a given volume of piperine was larger when the ultrasonic beam was applied, especially at high degrees of supercooling (beyond Tamman's maximum). The preparation was not specially purified. Danilov and Teverovskii [19] examined this effect on supercooled piperine, salol, and o-chloronitrobenzene; the material was contained in a flat-bottomed glass tube of internal diameter 6-7 mm. The temperature inside the filled tube was measured with two thermocouples, one at the center and one at the wall. The variation in temperature with height was recorded by moving the thermocouples. The tube was then filled with the purified material and was sealed off. The use of these calibrated tubes provided a means of determining the supercooling at the site where the first nucleus arose without the disturbance produced by a thermocouple. They concluded that the power levels used (2 W/cm^2 at 1 Mc) caused a large increase in the number of nuclei for high degrees of supercooling (20-40°) no matter whether the piperine was purified or not. Nuclei arose at impurity particles in unpurified salol; here insonation raised the temperature at which the first crystal appeared. Purified salol did not give nuclei at any supercooling. Impurities had little effect on o-chloronitrobenzene; spontaneous nucleation is the main effect here. The temperature of nucleation rose almost to the melting point as the intensity of the ultrasonic energy was increased.

In every case the rate of solidification was increased by insonation.

Berlaga [20] examined the effects on the linear rate of crystallization for supercooled salol and on the number of nuclei for betol. Thermal effects of the beam were eliminated as follows. The molten salol was poured into a tube with flat sides, whose ends were bent upwards at right angles; the tube was placed in the irradiation cell a short distance away from the crystal, and the cell was fed with a flow of oil, whose temperature at various points was measured with thermocouples. The tube was engraved with marks to provide a means of reading the distance traveled by the crystallization front. It was found that the linear rate was increased when the salol was supercooled to room temperature; the precise effect being governed by the intensity. In the next study [21], special attention was given to distinguishing the thermal effect of insonation from other effects; to this end the thermal effect, which reduces the linear rate of crystallization, was minimized by cooling the tube with a fast flow of oil. Of course, the effect could not be eliminated completely, so the residual effect was deduced from the times of transit of the front for equal distances with and without the field. The frequency was high (6 Mc), the source crystal being 2 x 25 x 30 mm; various crystals were used. A glass tube of oval section having flanged edges and a flat base was filled with the material and was placed with its flat base on the crystal, acoustic contact being provided by oil. It was found that the linear rate and the intensity increased together if the cooling was good; the rate could be ten or more times that obtained under ordinary conditions. Berlaga also found that the number of nuclei was increased in betol, which led to a considerable increase in the rate. The field also affected the appearance of the solid product (the frequency in this case was 1660 kc). The acceleration was explained on the assumption that the field separates the fresh nuclei from ones formed previously.

Mikhnevich and Dombrovskii [22] examined the effects of low-frequency (256 c/s) vibrations on the number of nuclei generated in a layer of betol about 0.2 mm thick; here the source was a maintained tuning fork, one of whose limbs was attached to the glass base. The tests were done at temperatures above the optimum crystallization temperature. In this case, the waves reduced the number of nuclei.

Belynskii [23] examined the effects on sodium thiosulfate; a glass tube was filled with the molten substance, one side of the tube being fitted with a metal rod. Acoustic vibrations were excited by rubbing the rod with leather coated with rosin; these caused the formation of a crystalline dust, which was distributed between the nodes and antinodes, the result being layers that grew together. Here the waves produced many nuclei.

Shablykin [24] used solutions of sulfur in dichlorethane with strong waves at 600 kc; the field reduced the temperature at which crystals began to appear by 2°, and the shape of the crystals was also affected. Most of the crystals were ones of the orthorhombic modification. Up to 95% of the crystals were deposited as aggregates in the absence of the field. The field altered the mean size of the crystals and also the limits to the range. The waves were considered to produce effects analogous to those of impurities, in that they altered the dispersion and the shape. The field increased the range of dispersion by a factor 5.

Mazhul' [25] supposed that cavitation, which gives rise to electric fields at the boundaries, could cause spontaneous nucleation; his experiments gave rise to the conclusions briefly summarized as follows: nucleation in the melt cannot be explained in terms of electric fields associated with cavitation; the sites of cavitation themselves act directly as nuclei, but cavitation cannot be taken as the sole cause of the effects, for weak fields (which do not produce cavitation) still displace the peak on the nucleation-rate curve to lower temperatures.

S. Ya. Sokolov [26] found that insonation produced more rapid crystallization when the solutions were seeded in the case of beet sugar and zinc sulfate.

I. T. Sokolov [27] found that the time for complete crystallization of supercooled water decreased as the supercooling or the ultrasonic intensity was increased. The liquid as a whole crystallized rapidly at 8-10 W/cm^2 on account of the production of large numbers of nuclei. The crystallization was less rapid at 4-8 W/cm^2; it still occurred at 2.5 W/cm^2, but then there was a delay of 4 min. There was not found to be any effect of frequency.

Danilov and Chedzhemov [28] used various supercooled liquids at 900 kc; they concluded that dispersal of the growing crystals and motion of the crystallites are the causes of the increase in growth rate. Insonation of alloys caused a large reduction in the primary grain size and ejection of the solid impurity particles. This ejection appeared to increase with the power input for uniform fields, even particles of lead were ejected from water at very high power levels. This effect could be used to remove nonmetallic inclusions from alloys. There was not found to be any power level at which ejection was replaced by mixing, even at very high levels, but mixing did occur if the field was not homogeneous. The rate of nucleation in the absence of the solid phase was affected in the case of salol, which is caused by effects on the energy of nucleation.

Günther and Zeil [29] examined the effects on the crystallization of glycerol and benzophenone; effects on the linear rate were distinguished from ones on nucleation by selecting substances whose kinematic parameters were widely separate on the Tamman temperature curve. The linear rate was found to increase with the hydrostatic pressure. Power adsorption (heating) affected the rate of crystallization of supercooled glycerol; the rate was reduced at small degrees of supercooling. The rate for benzophenone was always increased by insonation, probably on account of removal of the heat released at the interface between the phases. The difference between glycerol and benzophenone is associated with the difference in the latent heats; there is no temperature rise at the interface for glycerol.

Gas in the melt can cause a large rise in rate when the field is applied; this is considered to be caused by the turbulence associated with the release of the gas. The conditions for benzophenone for small degrees of supercooling can be made such that the grain size of the product alters over distances of $\lambda/2$ (λ is the wavelength of the standing waves).

Turner et al. [30] used frequencies of 8-20 kc at 100 W/cm^2 with various sugar solutions; the field was found to improve the homogeneity of crystals grown from seeds.

Fig. 7. Structure of zinc, ×100: a) Unirradiated; b) irradiated.

There are many papers on the effects of ultrasonic energy on metals; the idea that vibration during crystallization might improve the quality of the castings is not new, but modern ultrasonic techniques have enabled us to detect new effects.

S. Ya. Sokolov [31] applied high-frequency ultrasonic energy to solidifying metals (zinc, aluminum, tin) and observed effects on the structure. The metals were contained in a steel crucible and were treated at frequencies between 600 and 4500 kc. The metals were found to solidify more rapidly and to have very much altered microstructures, especially zinc; the grain size of the zinc was much reduced, and it was considered that the ultrasound facilitates the formation of a dendritic structure. The frequency was found to influence the crystallization time and the structure.

The time of crystallization for zinc at a given temperature was found to be reduced by 10% at 700 kc and by 35% at 3.7 Mc. The grain sizes obtained with 1200 and 700 kc were rather different.

Schmid and Ehret [32] used a 10-kc magnetostriction generator with melts of duralumin, silumin, cadmium, and antimony; the latter two crystallize with large grains under normal conditions, especially when the

a   b

Fig. 8. Effects of ultrasound on the columnar structure of zinc: a) Unirradiated; b) irradiated.

cooling is slow, but the ultrasound reduced the grain size greatly. It was considered that the field breaks up the grains while they are growing, with the results that the number of them is greatly increased. This smaller grain size results in new properties; for example, antimony becomes much less brittle, and the mean Brinell hardness increases from 34 to 52 kc/mm^2. The solidification time is not affected. Duralumin and silumin give similar results; the ultrasound produces a fine-grained structure. The Brinell hardness of duralumin is increased from 78 to 96 kg/mm^2, but silumin shows an increase from 38 to 39 kg/mm^2 only, which is within the limits of error of experiment. These workers also examined the cooling curves for these metals; antimony under irradiation showed no supercooling, but the cooling curves for the others were not affected by the ultrasound.

Schmid and Roll [33, 34] used Wood's metal in tests on the effects of frequency and intensity. The frequencies were 9 and 285 kc; the 9 kc was produced by a magnetostriction generator, and the higher frequency by a quartz plate. The two frequencies gave the same result at any given intensity, from which they concluded that friction between solid and melt is responsible for the reduction in grain size. Their calculations show that the frictional forces were comparable with the yield point; they took the view that S. Ya. Sokolov's results for zinc in a steel crucible do not demonstrate that ultrasound facilitates the growth of dendrites, for the melt behaves in strong fields as do other metals. They explain S. Ya. Sokolov's sections in terms of the presence of FeZn$_7$ in the form of needles, which were erroneously taken to be dendrites, for they found that the dendrites were broken up by the ultrasonics in their experiments.

Seemann[35] stated that it is to be expected that ultrasound will influence the formation of solid solutions.

Nomoto [36] found that ultrasound reduced the grain size in castings of bismuth and Wood's metal.

Kapustin [37] examined the effects of ultrasound (500 kc) on metals (zinc, duralumin, lead, Wood's metal) at intensities up to 0.3 W/cm^2; the mean time of complete solidification was much reduced, and the grain size was reduced, the results being highly reproducible. The effects on the grain size were especially pronounced for lead and zinc (Figs. 7-9) but were not great for duralumin, probably on account of the low intensity (Fig. 10). Zinc and duralumin showed increases in hardness (from 28 to 39 kg/mm^2 and from 69 to 76 kg/mm^2, respectively); aluminum and lead showed no change within the errors of experiment.

16

The Wood's metal had the composition 48% Bi, 26% Pb, 13% Sn, 13% Cd; melting point 62°C. It was necessary to reveal the structural components (e.g., cadmium) in the polished sections, which were etched with 50% HNO_3 in alcohol. Long needles were produced when Wood's metal solidified under normal conditions (Fig. 11a); Fig. 11b shows that the ultrasound (frequency 720 kc) eliminated the long needles of cadmium almost completely; only a few fragments remain visible.

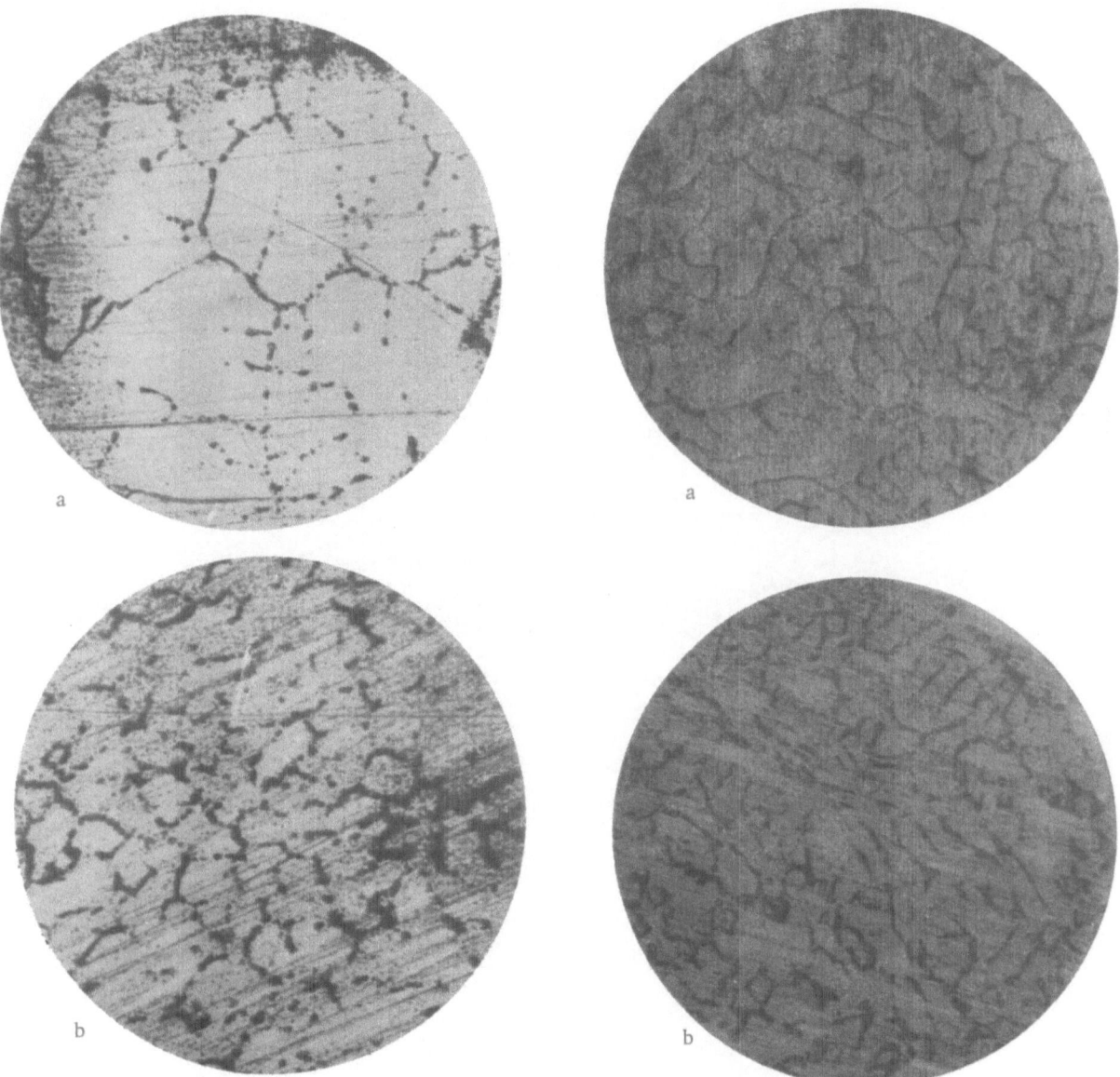

Fig. 9. Structure of lead, ×100: a) Unirradiated; b) irradiated.

Fig. 10. Structure of duralumin, ×100: a) Unirradiated; b) irradiated.

Gurevich et al. [38] used high-power ultrasound (18 kc) on heat-resisting steels and found that the linear dimensions of the grains were thereby reduced by factors of 3 to 5; the columnar crystals were broken up, the nonmetallic inclusions became uniformly distributed, and dendrite liquation was reduced. The mechanical properties (malleability) were also affected.

Teumin [39] examined the heat-resisting alloys ÉI530, Kh27, Kh25N20, N35KhMV, ÉI595, and also cast iron; the grain size was reduced by insonation and the dendritic structure was lost or substantially altered. The castings were 35-50 mm in diameter, up to 90 mm long, and up to 2 kg in weight. Some of the mechanical properties were altered; steel Kh27 showed the relative contraction increased by more than a factor 7, the rela-

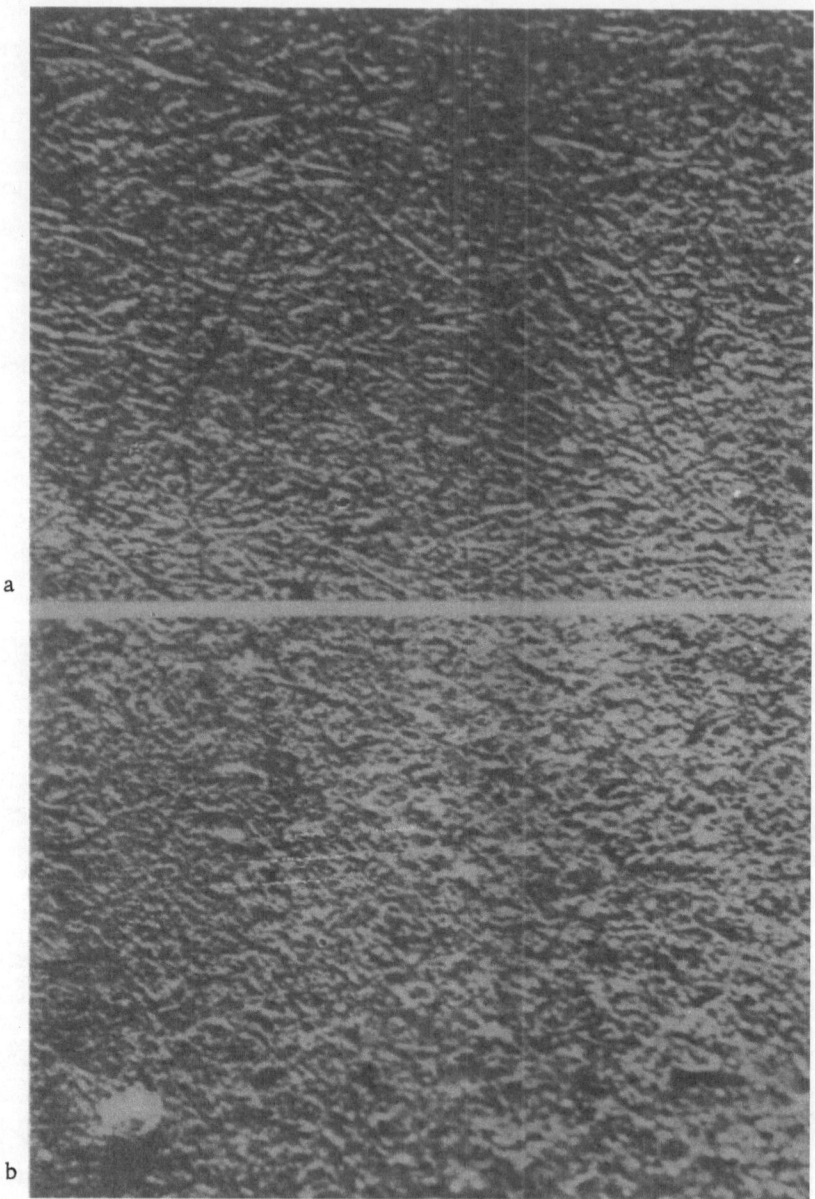

Fig. 11. Effect of ultrasound on the crystallization of an alloy of eutectic
type (× 50): a) Unirradiated; b) irradiated.

tive elongation by a factor 3, and the relative shear by a factor 2. These changes persisted after heat treatment but were then not so pronounced. For example, the impact viscosity of the irradiated material was increased by a factor 2.5, but the ratio fell to 1.5 after heat treatment. The malleability of cast Kh27 increased by a factor 4.5 for cold rolling after insonation, by a factor 3 after insonation and heat treatment, by a factor 1.15 for hot rolling (1200°C) after insonation, and by a factor in excess of 1.2 after insonation and heat treatment. (This parameter was measured as the degree of reduction that produced cracks, a laboratory rolling mill being used.) The conclusions were that insonation improves the structure, the mechanical properties, and the homogeneity of the casting. The chemical composition (including gases) was not affected.

Nikolaichik and Nikolaichik [40, 41] compared specimens of cast iron molded in the usual way and under the action of ultrasound; the source was a magnetostriction vibrator made up of strips of permendur.

The irradiated specimens had a homogeneous structure and the same hardness at all points; they were resistant to wear, had dense structures, and were free from gas cavities.

Rostoker and Berger [42] used various alloys; they concluded that the intensity rather than the frequency (range 60 to 900 kc) governs grain size. They considered that the following factors govern the degree of reduction in grain size in an alloy whose structure is mainly of eutectic type: 1) rate of cooling (low rates give the greatest reduction, especially when the volume is small), and 2) intensity, which has little increase in effect above a certain level.

Seemann and Menzel [43] used four magnetostriction generators simultaneously at 20 kc to irradiate molten duralumin; the products were more homogeneous and of smaller grain size. The crystals of eutectic also had a smaller structure. The resistance to compression differed from the usual value by 3 kg/mm^2.

Polotskii and Benieva [44] used frequencies of 200 c/s and 500 kc with chemically pure cadmium and aluminum, and also with Zn − Al − Cu antifriction alloy (84.2% Zn, 8.95% Al, 5.83% Cu, 1.02% other elements). Both frequencies were found to affect the primary crystallization structure of all these metals; the grain size of the cadmium was reduced, while the dendritic crystals of the β-phase were reduced in size and more evenly distributed at 200 c/s. The reduction in grain size was larger at 500 kc.

Siebers and Bulian [45] found that sonic and ultrasonic waves (50 c/s and 280 kc) facilitate eutectic crystallization in magnesium−aluminum alloys.

Uedzawa [46] showed that ultrasound (28 kc) affects the solidification of Al − Si alloys; it facilitates division of the primary crystals in supereutectic alloys and produces a smaller grain size. Further, the reduction in grain size is not proportional to the intensity; there is no further reduction beyond a certain intensity. He considered that frictional breakage is not the mechanism but rather the effects of the vibration on nucleation.

Sirota et al. [47] used ultrasound (18-18.5 kc) with alloys of aluminum with silicon (2.5 to 20% Si); the molten metal was poured into a heated mold screwed onto the concentrator of a magnetostriction source. An irradiated casting had a localized shrinkage cavity as well as cavities distributed throughout the volume; the structural components were more evenly distributed. The ultimate strength was increased on average by 11% and the strain at rupture by 75%. The impact viscosity and hardness were increased by 5 to 20%, the silicon content being the decisive factor.

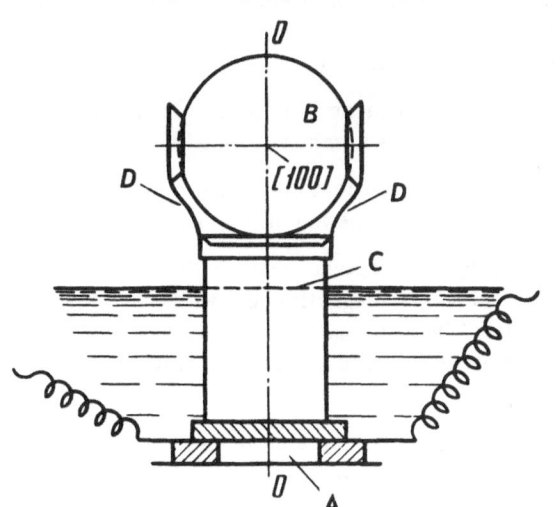

Fig. 12. Scheme of apparatus: A) Quartz plate (source); B) monocrystal; C) waveguide; D) holder.

2. Effects of Ultrasonic Energy on Solids

Ultrasound affects solids as well as solutions and melts; for example, the stress pattern in a monocrystal of thallium bromide−iodide can be seen to alter when high powers (4 W/cm^2 at 700 kc) are used [48]. The TlBr−TlI system is cubic; the material has the highest photoelastic constant known. Several tests were done with transparent crystals, which were observed in polarized light while exposed to the ultrasound. Figure 12 shows one way of placing the specimen B (monocrystal) on the quartz plate A (frequency 720 kc). The residual-stress pattern was photographed first with the field off, then with it on, and finally with the field off. Figure 13 shows that the field alters the distribution considerably; the isochromes vanish at once (these indicate the stress distribution) and the crystal appears uniformly light, with fluctuating colored areas. The original distribution is regained at once or after an interval (the intensity is decisive here) when the field is removed. The changes in the field are associated with the additional stresses. The effect can be used to measure the pressures set up.

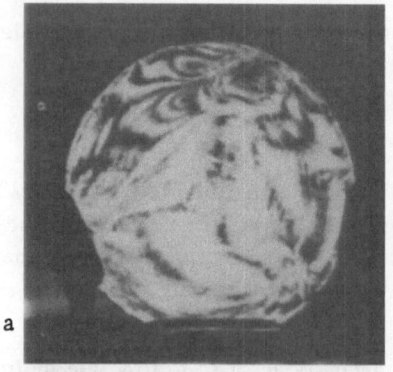

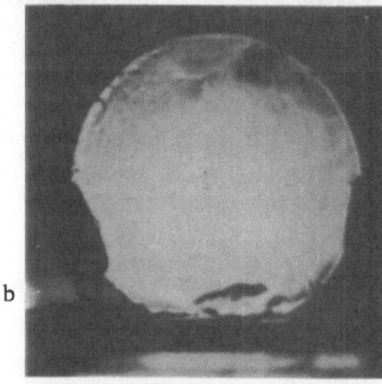

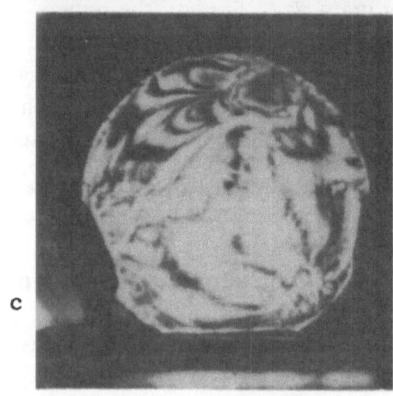

Fig. 13. Effects of ultrasound on the stress distribution in thallium bromide-iodide: a) Before; b) during insonation; c) after.

Ultrasound accelerates the conversion of white tin (tetragonal) to the grey form (cubic); it can also accelerate the reverse conversion above the transition temperature.

Ammonium nitrate and mercury iodide are very convenient materials for observing the effects of ultrasonics on polymorphic transitions. The transition points of the first are 125, 84, 32, and −16°. The cubic form is stable from 125° to the melting point; the trigonal form, between 84 and 125°; the α-orthorhombic form, between 32 and 84°; and the β-orthorhombic form between −16 and 32°. Several interesting effects have been observed at intensities of up to about 2 W/cm² at about 1 Mc. The transitions at 125, 84, and 32° were unaffected by the field at intensities up to 1 W/cm²; but at higher intensities the application of the field reversed the direction of motion of the boundary between the trigonal and cubic forms (temperature rising). The motion reversed as soon as the field was removed. The whole process could be repeated as often as desired. At a certain intensity, the field halted the motion, the boundary in the solid remaining in a fixed position for 60-80 sec; thereafter the motion continued. The same occurred at the 84° point, but not for the 32° one. No tests were made at the fourth point. Mechanical processes are the main ones governing the transition in this case [49]. Ultrasound also accelerates the conversion of yellow mercuric iodide (the less stable form) to the red form [37].

Kapustin and Popova's experiments with ferrites immersed in water showed that insonation at 30 kc greatly reduced the brittleness.

Hollmann and Bauch [50] found that ultrasound reduced the energy needed to reverse the magnetization of a nickel rod.

Schmid and Jetter [51] applied frequencies of 10 and 20 kc to nickel and detected an increase in magnetization, which itself increased with the intensity. The ultrasound also reduced the residual magnetization. The hysteresis loop became much narrower in the presence of the ultrasound.

Kapustin and Minskov found that an aluminum monocrystal showed an increased number of x-ray reflections after insonation at 30 kc for 1 hr at 0.3 W/cm².

Mahoux [52] has used the structural loosening produced by ultrasound to cause rapid enrichment of steel with nitrogen.

Much attention has been given to crystalline and amorphous polymers recently; the structures of these would indicate that insonation should cause interesting effects in them.

The surface of a crystalline body immersed in a liquid is altered by exposure to ultrasound; additional inhomogeneities are produced. X-ray analysis also reveals damage to the grains and disturbances in the structural orientation. Few studies have been made on the response of amorphous bodies to insonation.

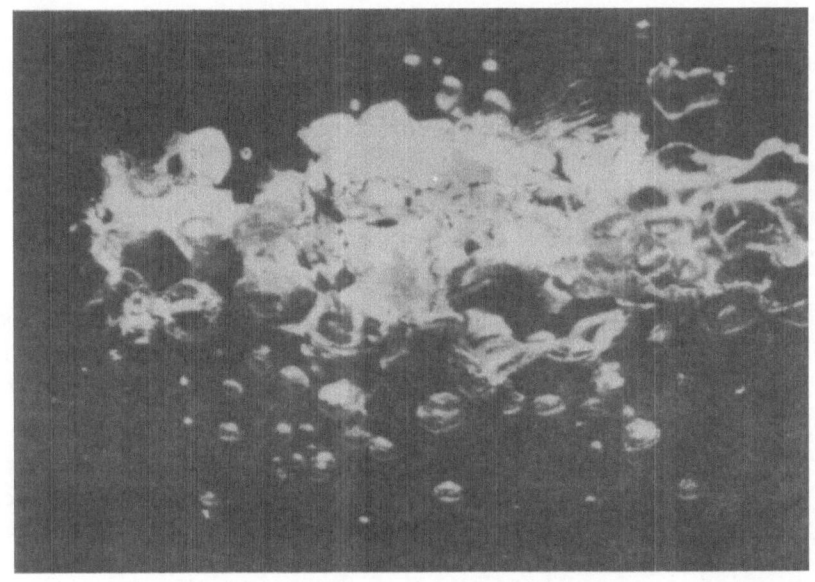

Fig. 14. Gas bubbles in a plexiglas plate.

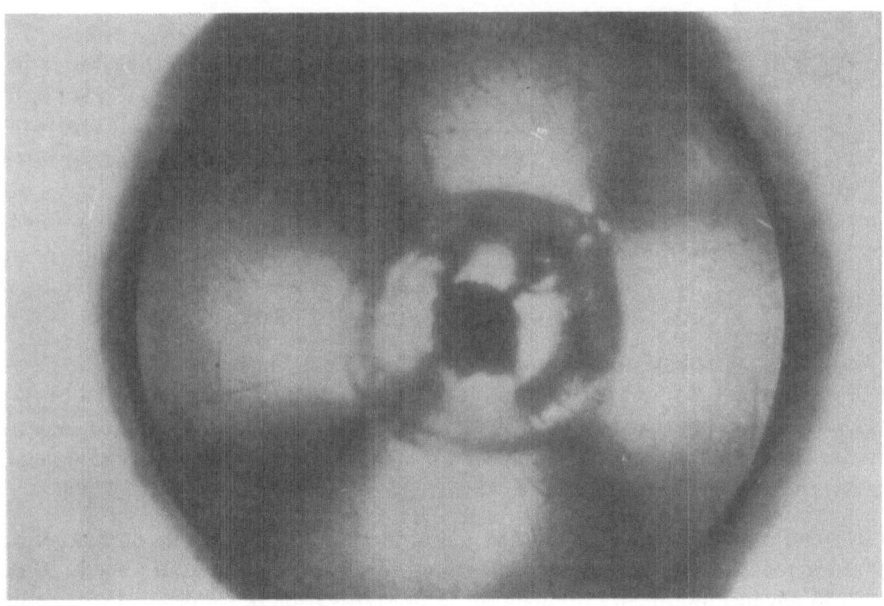

Fig. 15. Plexiglas plate with circular hole examined in polarized light after insonation (720 kc, 6 W/cm^2).

Here I report some observations on processes in amorphous polymers exposed to ultrasound [53].

In work with ultrasonic focusing devices, it was observed that the surface of a plexiglas lens handling a powerful beam soon became distorted and showed curious ripples. The extent and distribution of these are related to the intensity, frequency, and duration of action of the ultrasound.

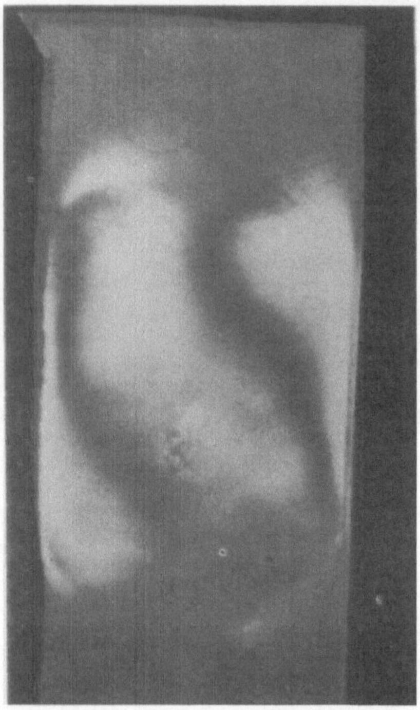

The source in this case was a flat quartz plate 60 mm in diameter and 4 mm thick. The radiated power in the oil was about 3 W/cm at 720 kc.

A plate of plexiglas of area 20 cm^2 and thickness 0.8 cm was fixed with one side in contact with the oil and was irradiated for a certain time, after which it was examined in polarized light.

The first effects visible to the eye appeared in 20 sec, and these spread throughout the plate within 2 min. The plate at the same time became very hot. The ripples were formed on both sides, but the area and height of these were larger on the side facing the oil; they also persisted when the beam was switched off and the plate had cooled down. Prolonged insonation gave rise to numerous gas bubbles under the ripples (Fig. 14), first on the side nearer the source and then throughout the plate. The surface of the ripples themselves remained smooth even on prolonged irradiation; it did not become rough.

The insonation produces internal effects visible in polarized light; the effects are also dependent on the stresses set up in the material during mechanical treatment. Figure 15 shows the effect of a hole 1.5 cm in diameter drilled in a plate subsequently irradiated for a long time. Figure 16 shows a similar plate not given mechanical treatment; the plate showed no stresses before it was irradiated, so the exposure altered the structure of the polymer. Simple heating in oil does not produce this effect; nor does exposure to low intensities (0.3 W/cm^2 at 20 kc).

Fig. 16. Plexiglas plate not previously given mechanical treatment after insonation (720 kc, 6 W/cm^2).

This short survey shows that the effects are very varied and of considerable interest from the physical and physicochemical points of view. Such studies have thrown much light on nucleation and growth of new phases in the presence of ultrasonic fields, but the kinetics of these effects have not yet been examined in detail. The study of interactions between ultrasound and matter is impeded by the lack of reliable methods of measuring ultrasonic power, which is especially serious in relation to work with metals.

There have been substantial advances in the practical use of ultrasonics [54], but we do not yet have a complete understanding of the mechanism whereby the ultrasonic energy acts on materials. Fresh experimental studies are needed in order to advance the theory of processes of crystallization (dissolution) in ultrasonic fields.

CHAPTER III

CHAPTER III

Crystallization Processes of Organic Compounds

1. Thin Films

The glass waveguide described above (Fig. 1) is used to transport the energy to the film, which is observed under the polarizing microscope at a magnification of 16 to 48. The frequencies we have used range from 720 kc to 6 Mc, with power fluxes of about 1 W/cm^2.

A variety of growth structures may occur in the absence of the field (zoned, spherulitic, screw, and so on). For example, a slide bearing a supercooled drop of thymol is observed for the response to a seed crystal on the end of a needle and touching the glass at one point. A group of crystals starts to grow slowly from the point of contact. The application of the field alters the picture; there is a vigorous production of new centers at the interface between the phases. The new crystals are too small to see at the instant when they are formed; later, when they are no longer disturbed by the field, they grow to considerable size. Figure 17a shows the initial stage of this process, while Fig. 17b shows the grown crystals. The number of crystals increases rapidly, and the whole preparation becomes crystalline in a few seconds. Fresh nuclei are generated especially readily in regions where the vibrations are strongest. These effects have been observed with a number of supercooled substances.

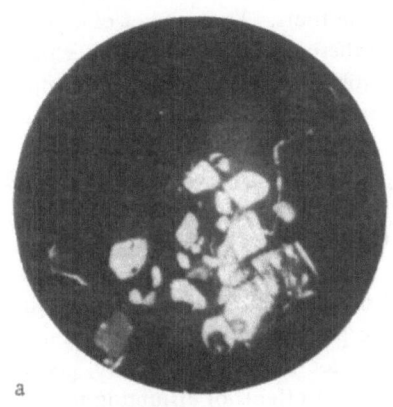
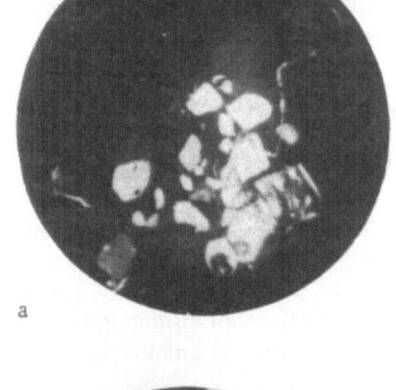

a

b

Fig. 17. Separation of fresh nuclei at the interface between crystal and melt in response to insonation: a) Initial stage; b) after prolonged irradiation.

The ultimate structure when the intensity is low consists of some regions having crystals of the usual size together with others of small grain size.

The following tests illustrate the formation of fresh nuclei more clearly. A thymol monocrystal was suspended inside a glass tube 200 mm long about 10-12 mm from the sides and near the middle. Water was admitted carefully to give a column about 100 mm long under the crystal; then the closed end was inserted in the ultrasonic fountain in the oil and was left there for 5-10 sec. A drop of water from the surface was then examined under the microscope; this was found to contain many small crystals of various shapes and sizes.

In a second experiment, a small crystal of thymol was fixed in a wooden holder and was immersed in water in a tube; then liquid thymol cooled to room temperature was carefully run in (the melting point of thymol is 51.58°). The thymol floated on the water, the interface being clearly visible; no crystals arose even after a long time, but crystallization began almost instantly when the field was applied. The reason is that the field broke off minute particles from the crystal in the water, and the wave pressure raised these to the interface, whereupon the supercooled melt began to crystallize. (Previous tests had shown that insonation of the thymol alone did not initiate crystallization.)

In a third experiment, the material was placed in a thermostat (a closed brass vessel with double walls; see description above) and was allowed to stand for 10 min after the set temperature had been reached; then a seed was inserted. The growing crystals were observed with a microscope; the ultrasonic field was applied when the growth front was about half-way across the field of view.

Fig. 18. Thymol allowed to crystallize under a coverslip, × 24; on the right, structure formed before the specimen was irradiated; on the left, structure formed afterwards.

Studies on melts [55] have shown that ultrasound reduces the time needed for complete crystallization by a very large factor; any substance that allows of considerable supercooling (thymol, salol, piperonal, benzophenone) always gives rise to a mass of crystals filling the space if a field of sufficient intensity is applied. Fresh crystals are produced not merely by breakaway from the existing growing surface but also within the body of the melt. The number of crystals continuously increases at first; then these crystals grow together into clumps. Nuclei can be generated from a seed at a variety of temperatures and intensities. Only o-chloronitrobenzene shows spontaneous crystallization (without a seed) in response to the field at moderate intensities. Some photomicrographs illustrate the behavior of thymol and other substances crystallizing with or without the field applied. Figure 18 shows thymol crystallizing under a coverslip; a coarse-grained structure is formed before the field is applied, and a fine-grained one afterwards. Figure 19 shows o-chloronitrobenzene with and without insonation; the structure produced by the field resembles the martensite structure of quenched steel. Menthol (Fig. 20) gives similar effects. Figure 21 illustrates the effects of disrupting the branches of dendrites occurring in a preparation of ammonium chloride.

Figures 22 and 23 show the effects of ultrasound on the crystallization of condensed milk and of molten sulfur.

In all, we have examined some 30 different substances; the results, in combination with published information, have enabled us to deduce the mechanisms whereby the crystalline phase is rendered more homogeneous and of smaller grain size. The mechanism is often simple, in that the field breaks off microcrystals, which act as fresh nuclei. The effect occurs over a wide frequency range (200 c/s to 9 Mc) provided that the intensity is sufficient.

These results are of interest as throwing light on the mechanisms of crystallization under very different conditions, e.g., in the presence of small quantities of impurity. The decrease in grain size is always associated with increase in the number of nuclei, no matter how produced (preservation by surfactants, production of new ones by ultrasonics).

The above method of observation has also enabled us to show that the continuous action of ultrasonic energy has a very pronounced effect on the growth kinetics of the crystalline phase at many different frequencies, intensities, and degrees of supercooling.

The numerous fresh nuclei alter the rate of crystallization and so provide a means of controlling phase transitions; it gives us a

Fig. 19. Structure of o-chloronitrobenzene, ×24: a) Little supercooling, no ultrasound; b) ultrasound applied.

means of displacing the maximum on Tamman's curve from its usual position and so enables us to produce materials of various grain sizes.

2. Crystallization in Bulk Material

Pronounced dispersion usually occurs at the interface between solid and liquid in the case of a thin film; it is of interest to see how the process is affected when a bulk sample is used.

Fig. 20. Structure of menthol at low degrees of supercooling, × 24: a) No ultrasound; b) irradiated.

Fig. 21. Ammonium chloride preparation: a) Before insonation; b) after insonation.

A special type of tube is used for bulk studies; this is made of quartz, has double walls, is 300 mm high and 5 mm in diameter, and has a thin base. The tube has a scale glued to it, which is used in recording the position of the phase boundary .

The outer jacket is fed with water at the required temperature, which enables one to keep the material superheated or supercooled at any point with an error of 0.1°C. Figure 24 shows the rate of crystallization of thymol as a function of temperature for melts not exposed to ultrasound.

The ultrasound was used at various intensities with various degrees of supercooling; the results [37] illustrate the effect of the waves on the rate of the transition in the bulk substance.

Here the source was a Curie-cut quartz plate 60 mm in diameter and 8 mm thick, which was operated at 720 kc at voltages up to 12 kV. The intensity within the melt was not measured; the voltage applied to the plate determines the intensity under fixed conditions. The beam was admitted through the thin base, which was brought into contact with the plate via a layer of transformer oil. Crystallization was initiated by introducing a seed. The shape of the interface altered as it moved when the material (thymol) was not irradiated. The

Fig. 22. Preparation of condensed milk: a) Before insonation; b) after insonation.

Fig. 23. Preparation of sulfur: a) Before insonation; b) after insonation.

time required for complete crystallization was about 30 min for a supercooling of 30°. The product was very uneven in structure when viewed in cross section. The field produced an even interface of fixed form (Fig. 25), though here the solid grew throughout the volume rather than at fixed points. Complete crystallization required only a few seconds, and the structure was the same in all parts. It was found that the rate of crystallization was governed by the intensity; the minimum electric field that produced organized movement of the interface was 0.6 kV/mm. Lower electric fields resulted in delayed crystallization and inhomogeneous structure, and there was no appreciable affect at all below 0.07 kV/mm. The corresponding intensity represents a threshold value for the effect of the field on the interface. In every case, removal of the field caused the rapid crystallization to stop in 4-5 sec; a cloud of small crystals settled onto the interface, and these enlarged as the growth continued (Fig. 26). The crystals were not visible at first at the interface; they became visible only after they had enlarged somewhat. The number of crystals varied with the intensity. These crystals were displaced from the interface if the field was applied again. The process when the field is applied continuously is thus that the smallest crystals break away from the boundary, grow larger, and tend to settle out, but are then forced upwards again. This process, in conjunction with the

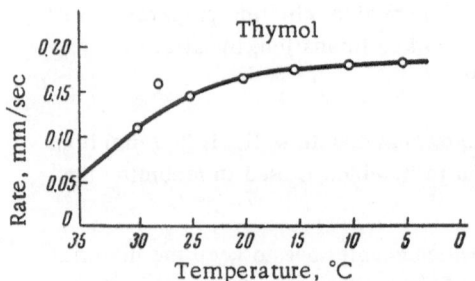

Fig. 24. Relation of rate of phase transition in thymol to degree of supercooling.

26

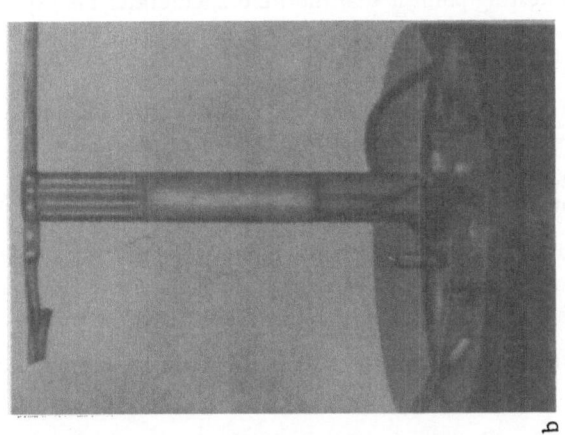

Fig. 25. Growth of a crystalline phase under the action of ultrasonic energy: a) Start of insonation; b) process developing; c) complete crystallization of thymol (time of insonation 8 sec).

27

limited size of the tube, causes a rapid motion of the phase boundary towards the radiation source. The ultrasound causes the minute crystals to break away from the interface and form a dispersed phase in the melt. This dispersion is accompanied by crystallization and coagulation; the latter tends to reduce the surface area, but this is more than balanced by the increase resulting from continuing dispersion.

Fig. 26. Cloud of crystals sinking after field has been removed; the gap in the solid was caused by a period when the field was off.

The result of these various processes is that the rate is increased considerably.

An ordinary projection system may be used to observe the effects when the ultrasonic intensity is low. Near the interface one sees a small number of crystals of various sizes, which oscillate in the field of view; as they reach a certain size, they are displaced towards the interface. Coalescence and dispersal can be seen to occur simultaneously; the way in which the crystalline structure is built up can be seen if the intensity is low. It is also found that crystallization is accelerated if the beam is injected from the side along the interface.

An interesting point is that the field accelerates the crystallization although it heats the melt somewhat (this is not entirely suppressed by the thermostat).

Figure 27 shows the rate as a function of voltage and supercooling; these curves give rise to the following general conclusions. The rate tends to a limit as the voltage is increased at fixed frequency and supercooling; there is also a lower limit (threshold value), below which the ultrasound has little effect.

The rate as a function of temperature has a maximum near 30°; the rate under ordinary conditions is 0.15 mm/sec, but the ultrasound increases the rate to 25 mm/sec.

The amount of material crystallizing in unit time may be calculated from the diameter of the vessel and the time of crystallization.

The processes that occur in such cases may be described as follows. The following two stages occur. In the first, the field produces a large number of fresh nuclei; in the second, the resulting crystals coalesce. The final homogeneous fine-grained product is a result of the lack of space for free growth.

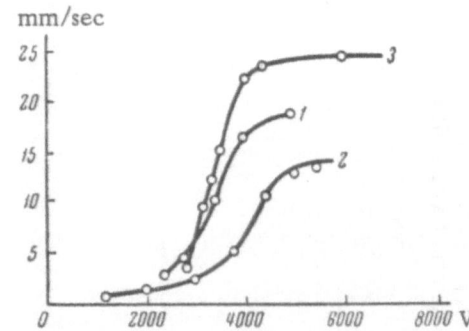

Fig. 27. Relation of growth rate for thymol to voltage applied to quartz plate for melt temperatures of: 1) 22°; 2) 35°; 3) 30°.

The crystals that break away from the growth front cannot become very large, for they themselves are soon broken up to form fresh nuclei.

In strong fields (2 kV/mm) and at moderate degrees of supercooling (25°), the rate of crystallization increases enormously as soon as seed crystals are introduced. The seeding may be done with a monocrystal or with powdered material.

Weak fields do not produce nearly so many fresh crystals, and the product is then of larger grain size.

There is an upper limit to the response to increase in ultrasonic intensity (voltage applied to the crystal) as regards crystallization rate and grain size; this upper limit is probably a result of the release of much latent heat, which retards the production of fresh nuclei. The lower limit represents the level at which cavitation is absent and which produces only slight vibration of the walls.

The small crystals do not break away from the interface when the field is applied only briefly and the supercooling is 40-43°; the rate of crystallization is not altered even when the field is applied for a long time. In this case, the solid grows by gradual deposition of layers, and the rate is little affected by the field.

An interesting effect occurs here. When the interface has traveled part of the way, a crystalline phase arises some 35-40 mm away from it, and this then grows independently. The effect is readily reproduced; it is difficult to imagine that it can be caused by the passage of crystals from the interface to remote parts of the melt, for the liquid is very viscous.

A crystalline seed introduced into a melt supercooled by 5-6° is rapidly dispersed by a strong ultrasonic field, the result being a cloud of crystals dispersed throughout the melt. This effect is not critically dependent on the intensity. It is impossible to estimate the crystallization rate under these conditions, for there is no sharp boundary between the phases.

The following effects are observed when the melt is supercooled by about 30°.

A typical fine-grained structure is formed when the potential applied to the quartz plate is 6-8 kV.

Medium voltages (about 3 kV) produce a rather lower mean rate (about 5 mm/sec), the structure being coarse-grained. The transition shows an induction period, and it stops not far from the base, because the intensity is highest there. The growth at the walls is much more rapid than that within the volume, on account of the vibration of the walls. The interface is of complex and variable shape, so it is difficult to record the times needed in the deduction of the rate, though the rate of crystallization definitely increases as the source is approached, with the result that some of the melt below the interface ultimately crystallizes before parts higher up, although they started earlier. The reason for this is that small crystals fall into the lower part and cause very rapid crystallization, for the intensity in that part is higher.

Low voltages (1-1.5 kV) give a highly inhomogeneous product, though the field does continue to affect the nucleation down to 600 V (150 V/mm in the plate). The transmission of the waves along the walls of the tube allows the vibration to affect the crystallization down to low voltages.

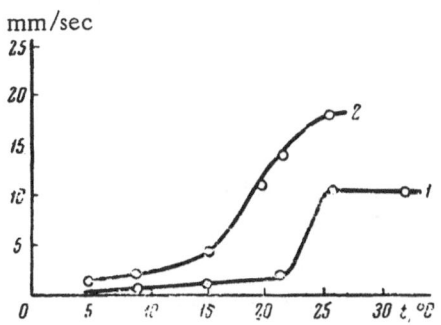

Fig. 28. Rate of growth of the crystalline phase as a function of temperature for potentials applied to the quartz plate of: 1) 2400 V; 2) 5200 V.

Figure 28 illustrates the effects of temperature on the crystallization rate, which clearly reaches a limit at about 25°C. This rate is nearly doubled when the potential on the crystal is raised from 2400 to 5200 V; the rate is then more than 100 times that observed under ordinary conditions.

This effect occurs even if the ultrasound is admitted other than through the bottom of a vertical tube. Further, similar effects occur for tubes of any shape. For example, we have used a bent tube in these experiments with thymol, the ultrasound being admitted through the base. Here again the crystallization rate gradually increased as the source was approached.

Few fresh nuclei are produced if the ultrasound is admitted from the side at a considerable vertical distance from the interface; the rate of crystallization increases as the point of admission is brought closer, and it becomes maximal when the beam passes close to the interface.

In this way, the rate of crystallization can be controlled by means of the distance from the interface for a fixed field intensity.

A further effect is the virtually complete elimination of gas from the melt, which occurs at medium and (especially) high intensities.

3. Structure of Castings. Mechanical Properties

It is simple to examine the structures of materials cast with or without the field present [37]. Organic materials may be cut with a silk thread kept moistened with alcohol; the cut surface is carefully polished with a soft cloth and is then photographed. Figure 29 shows the result for cast thymol specimens cut lengthwise; the first specimen (Fig. 29a) was grown under ordinary conditions, while the second (Fig. 29b) was grown in the presence of an ultrasonic field (intensity about 6 W/cm^2). The temperature was 30° in both cases. The first

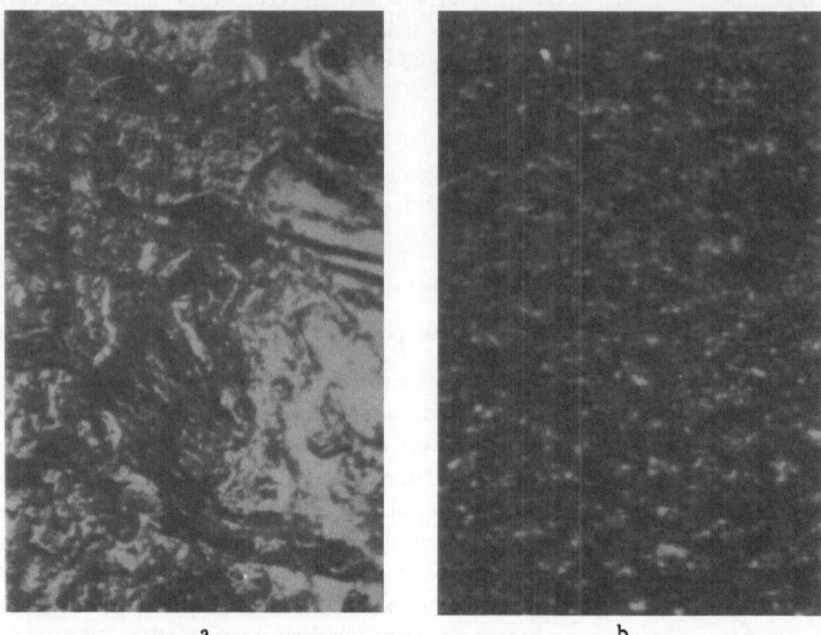

a   b

Fig. 29. Lengthwise sections of specimens of thymol crystallized: a) Under ordinary conditions; b) in the presence of ultrasound (the transverse lines represent flaws caused by the cutting).

has a highly inhomogeneous structure contains many gas bubbles and is very brittle, whereas the second has a homogeneous structure free from bubbles and is very strong.

Examination of such specimens in lengthwise and transverse sections showed that the homogeneity, taken over the specimen generally, was best at 6200 V and fell off rapidly as the voltage was reduced.

These specimens were also examined for their mechanical properties.

Tests in compression were made with an Amsler machine on 50 identical specimens; the average rupture stress for specimens grown in the ultrasonic field was 50 kg/mm^2, whereas ones grown under ordinary conditions ruptured at 30 kg/mm^2. The results were highly reproducible. Similar effects were observed for metals crystallized in ultrasonic fields.

4. Effects of Ultrasonic Frequency on Crystallization

There is a tendency to assign the main role to the intensity of the ultrasound and to consider that the frequency has no marked effect on crystallization.

On the other hand, dispersal, emulsification, depolymerization, and other such processes are notably dependent on the frequency, the effects presumably being related to the variation in cavitation with frequency. This aspect of the matter is extremely important to the practical use of ultrasonics. Means of suppressing cavitation have now been developed, but relatively little is known about the conditions under which the effect is maximal.

Rzhevkin and Ostrovskii [56] were the first to observe that the dispersal of a metal may be frequency-dependent.

I [37] have used frequencies of 700 kc and 2 Mc at about 2 W/cm^2 in studies of effects on grain size and degree of homogenization.

Insonation of crystallizing melts at 700 kc gave more homogeneous structures and smaller grain size.

Bagdasarov [57] made a detailed study of cavitation effects in crystallization for a fixed intensity of 0.25 W/cm^2 at 30, 717, 5760, and 9600 kc. The change in grain size was determined from the size and number of

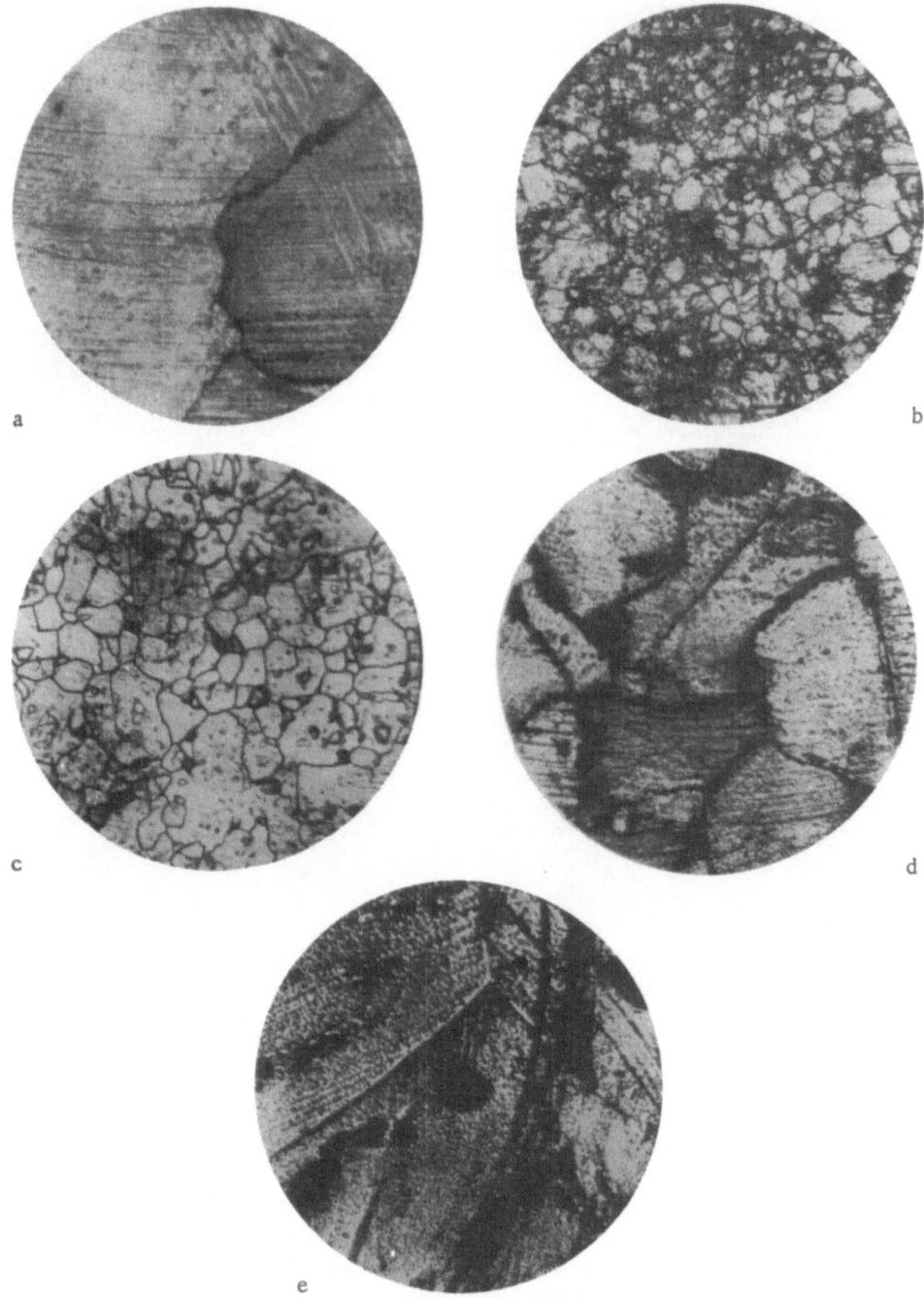

Fig. 30. Relation of grain size to frequency of ultrasound for cut surfaces of thymol, × 70: a) No ultrasound; b) 30 kc; c) 717 kc; d) 5760 kc; e) 9600 kc.

the grains in unit area of a cut section. The reason for the comparatively low intensity was that it was not possible to obtain high intensities within the volume of the melt at 9.6 Mc. The experiments were done as follows.

A tube containing crystalline thymol was placed in a thermostat; the material was melted, was heated to 65°, and then was cooled to 8°. The intensity was measured, and then a seed was inserted to initiate crystallization. The crystallized products were cut into slices, which were photographed to determine the number and size

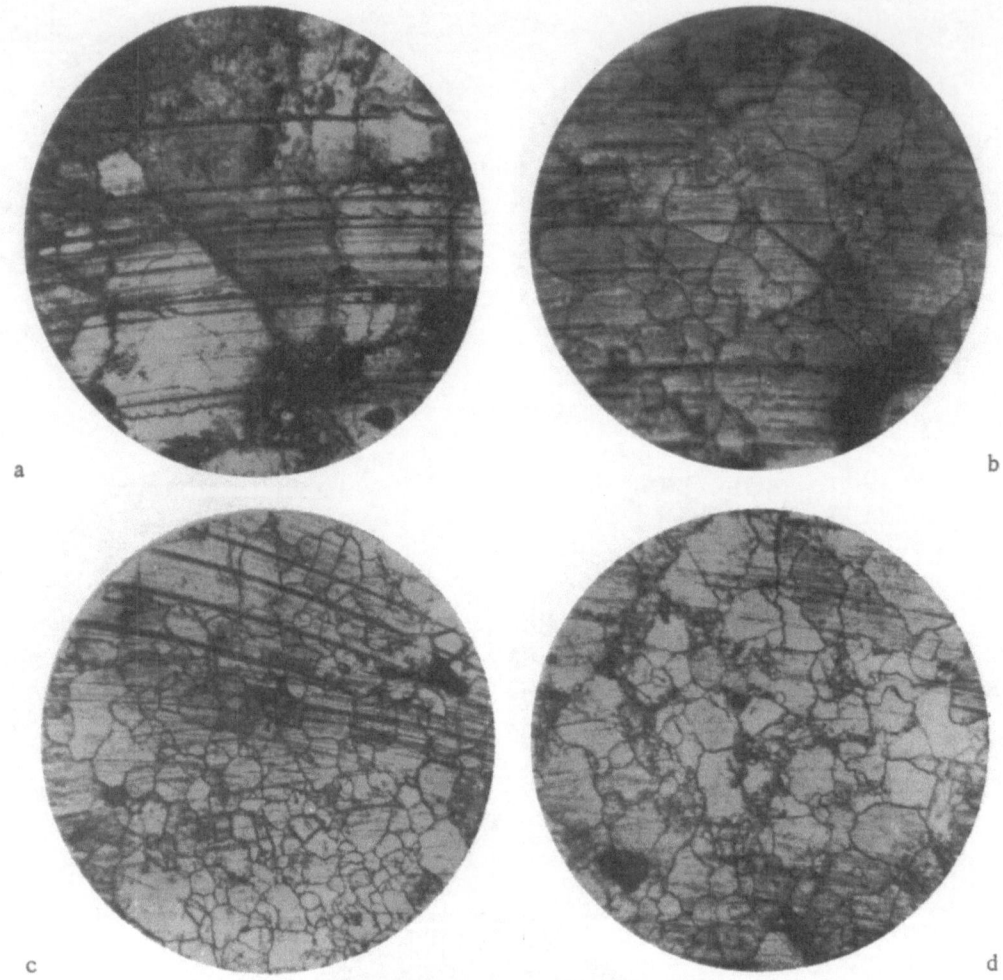

Fig. 31. Thymol crystallized at 8° under various external pressures (atm): a) 0.22; b) 0.49; c) 0.76; d) 1.03; e) 2.50; f) 4.50; g) 5.50; h) 6.00; × 70.

of the grains in 1 cm². Figure 30 shows sections made in this way; the grain size is clearly smallest for 30 kc; it is still fairly small at 717 kc, but there is virtually no effect at the higher frequencies. Bagdasarov ascribes the effect of frequency to variation in cavitation. Tests were made with conditions designed to vary the extent of cavitation. Hydrostatic pressure was used as having a pronounced effect; this was applied by means of compressed gas above the melt, reduced pressures being provided by a rotary pump. The material (thymol) was irradiated at 717 kc and 0.4 W/cm² (Fig. 31); there was a clear variation in grain size with pressure, the maximum reduction in grain size occurring near atmospheric pressure. There was very little effect on the grain size above 4 atm (Fig. 31). The conclusion drawn was that the frequency dependence of the grain size is governed by the frequency dependence of the cavitation.

5. Effects of Ultrasonics on Columnar Crystallization

A casting showing columnar crystallization usually has several zones differing in structure. The outermost zone consists mainly of columnar crystals (ones that have grown only at right angles to the walls of the

32

mold). Competition during growth causes the number of crystals to decrease inwards, the size of each simultaneously increasing; the largest crystals are found near the center, if spontaneous crystallization cannot occur.

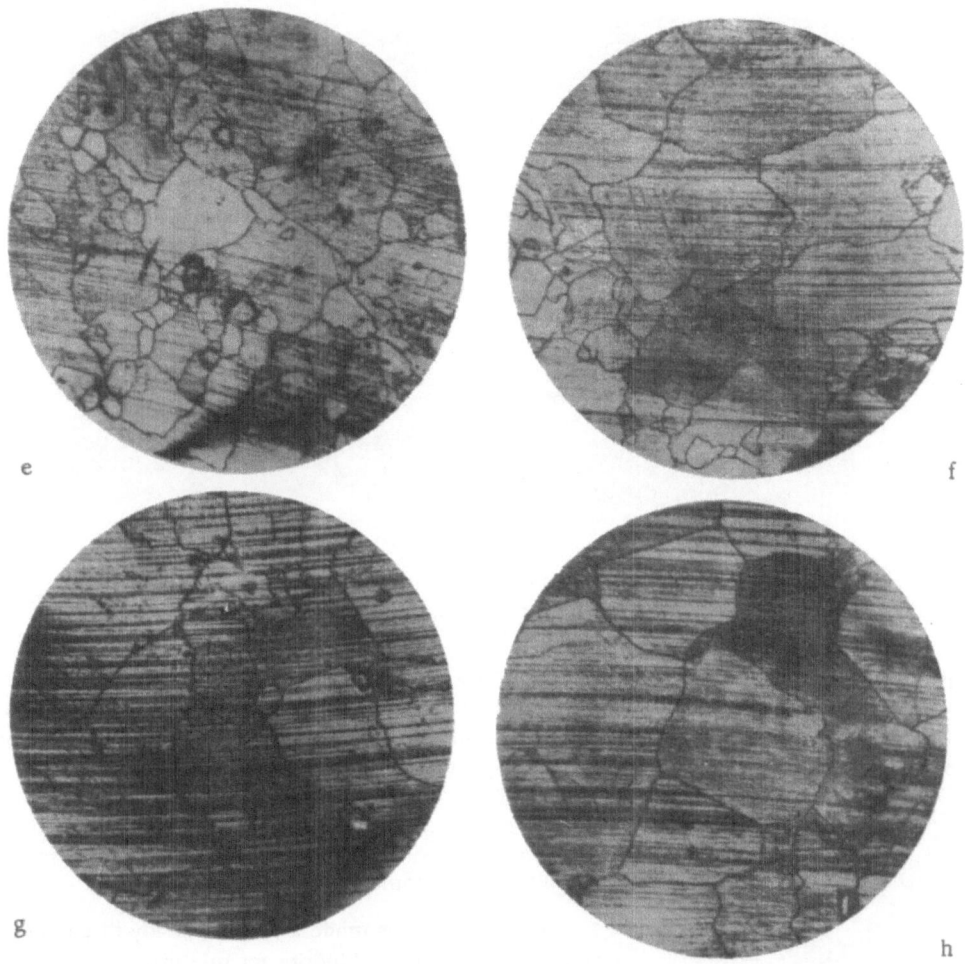

e   f

g   h

Fig. 31 (continued).

Columnar crystallization occurs under natural conditions in geodes and veins, in tubes encrusted with salts in chemical processes, and so on. The effect is known as orthotropism; it is of practical and theoretical interest because it is common to find that a columnar casting will break during rolling, on account of poor bonding between the oriented crystals.

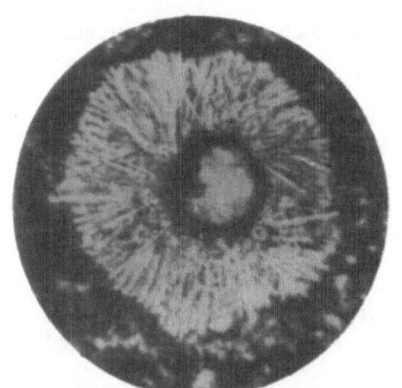

Fig. 32. Effect of ultrasound on the columnar structure of thymol; × 24.

Ultrasonic irradiation largely or completely eliminates the columnar zone [37]; Shubnikov and Lemmlein's method [58] was used to observe the growth of the columnar structure. Here a drop of the melted material is placed on a glass slide having a hole about 0.05 mm in diameter and is covered with a coverslip. A seed is inserted through the hole when the preparation has cooled to room temperature. The process is observed under the microscope, the product being a spherulite (cylindrite).

Figure 32 illustrates the effect of ultrasonic treatment; in the center we have the seed, then a columnar zone, and outside this a fine-grained zone resulting from the application of ultrasound.

Further experiments were made with a glass dish 40 mm in diameter and 60 mm deep; this was coated on the inside with solid

thymol and then was placed into a close-fitted double-walled metal jacket supplied with water. Thymol at 48°
was poured into the dish and was left until crystallization was complete. Similar experiments were done with

a   b

Fig. 33. Lengthwise sections of specimens of thymol: a) Grown without
insonation; b) grown under insonation.

the dish exposed to ultrasound. The solidified specimens were extracted and were cut along the axis with silk
moistened with water containing a little alcohol. Insonation completely suppressed the columnar zone (Fig. 33).
The results were highly reproducible provided that the intensity was not less than 3 W/cm^2.

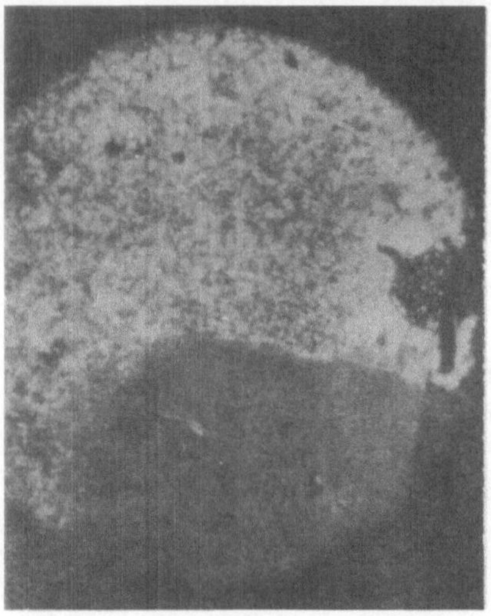

6. Crystallization of Mixtures of Eutectic Type

These show some interesting effects [37].

Bochvar [59] has made a detailed study of eutectic crystallization and crystallization kinetics for mixtures of organic substances; he found that eutectic crystallization proper begins when a crystal of the one substance meets one of the other substance, and that the rate of eutectic crystallization is higher than the rate for either substance alone in a liquid of eutectic composition. The shape of the crystals is dependent on the degree of supercooling.

I have used the same materials as Bochvar (azobenzene and piperonal).

My experiments showed that an ultrasonic field applied to a thin layer of the material causes crystals of azobenzene and piperonal to break away, which brings together crystals of the two species and so accelerates the onset of eutectic crystallization; moreover, the form taken by the eutectic (spherulitic at high degrees of supercooling) is thereby altered.

Fig. 34. Change in appearance of eutectic in response to field (× 24).

Figure 34 shows this effect of ultrasound; much of the preparation is clearly of small grain size.

7. Production of Zoned Structures

The production of zoned crystals is of interest in that the size and mutual orientation of the crystals are important to the physical properties of the substance. Ultrasonic studies indicate new ways of producing zoned crystal structures [60].

34

A zoned structure can arise from the action of any factor that influences the structure and that varies periodically.

Dust particles, gas bubbles, and so on take up positions at the nodes in a standing wave; the effect has not been examined in detail for supercooled melts or solutions, nor, so far as I am aware, has any attempt been made to use it in order to produce a polycrystalline aggregate with a periodic structure.

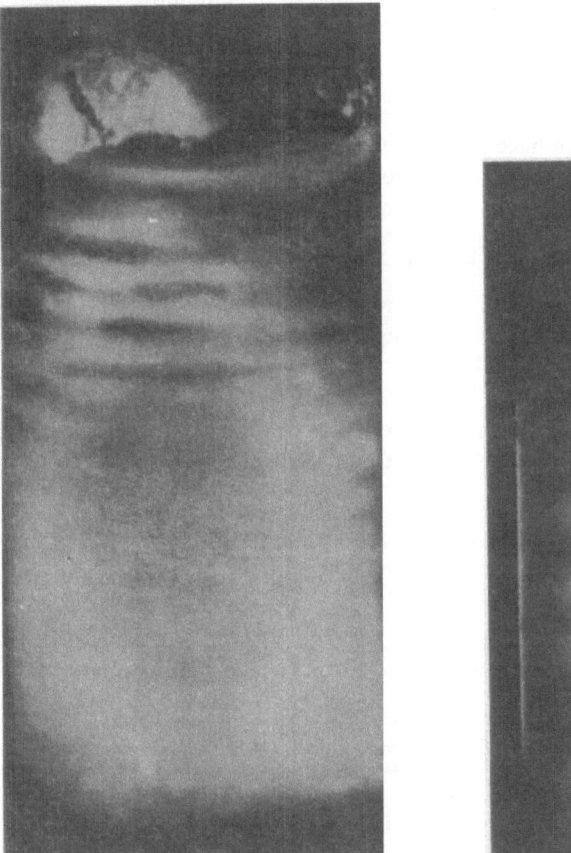

Fig. 35. Zoned structures in Rochelle salt (left) and o-chloronitrobenzene (right).

A glass tube 12 mm in diameter and 120 mm high was filled with the melt and was placed above an ultrasonic radiator. The power level (about 5 W/cm^2 at 720 kc) was measured calorimetrically.

Rochelle salt gave the most clear-cut results; treatment of a specimen at 70° gave rise to thin films separated by half the wavelength after 10-12 sec. These nodal planes were still visible when the melt had solidified, though they were not so clear. Figure 35 (left) shows the zoned structure produced in this way at 720 kc.

Microscopic examination showed that the structure at the antinodes was fine-grained, the crystals being randomly oriented; the nodal planes had larger crystals, also randomly oriented.

The distance between the nodal planes is governed by the frequency.

A zoned structure can be produced in some cases by applying the field periodically.

The speed taken up by the particles torn off by the field is dependent on the viscosity of the melt when the field is turned off; this speed may be larger or smaller than the speed of the phase interface (also in the absence of the field). It is possible for the interface to outrun the sinking particles, as in the case of o-chloronitrobenzene. Here the ultrasonic field merely breaks up the needle crystals, and this process continues while the interface is exposed to the ultrasound. The zones that are formed while the field is on have a fine-grained structure, the small needles having no regular orientation. Larger crystals and lower rates occur when the field is absent.

This provides a means of producing zones of controlled width and repeat distance.

Figure 35 (right) shows a zoned structure produced in o-chloronitrobenzene by periodically applying the field.

8. Effects of Vibration of the Walls of the Vessel

Wall-vibration effects have been studied many times; for example, Polotskii [61] remarks on them. I made a study of these effects in the crystallization of organic compounds in 1951. In 1956, Bagdasarov obtained some novel results illustrating the major effects of all vibration on crystallization.

The feature of importance here is that the vibration facilitates nucleation.

It is inconvenient to introduce the field through the bottom of the vessel when the volume is large, so use of wall vibrations is a convenient alternative; moreover, it can be more effective.

Measurements were made at 20 and 700 kc, with intensities of 0.3 and 5 W/cm^2.

It was found that crystallization in capillary tubes of diameter about 1 mm was accelerated when the lower end of the tube was free from melt and the ultrasound reached the lower end of the melt only through the walls.

Table 1 gives the results from the tests at 700 kc and about 5 W/cm^2.

Table 1

Length of column crystallized, mm		Time, min	Notes
no ultrasound	ultrasound		
30	60	5	In upper part of tube
30	140	5	In middle part
30	190	5	In lower part

The effects on a thin layer of material were observed by filling a wide tube with supercooled benzophenone, which was then emptied again to leave the walls coated with a thin film. The material began to crystallize when a seed was inserted. Application of the ultrasound (720 kc) to the walls gave rise to a pattern of nodes and antinodes, which remained clearly visible when the material had crystallized (Fig. 36). The crystals were localized around the nodes. No cavitation occurred under these conditions; only the vibration of the walls affected the crystallization.

Tests were also done at 20 kc and about 0.3 W/cm^2.

A glass beaker 50 mm in diameter was filled with melted benzophenone; a seed was inserted, and the vessel was placed on a magnetostriction generator immersed in water. The layer of melt was 80 mm deep. The floating seed gave rise to many small particles in response to the insonation; surface tension rapidly drew these away to the sides. The crystallization then propagated along the wall and across the bottom; the middle part of the material remained liquid for a long time. The crystals at the sides gave rise to small particles in response to the vibration; these soon filled the volume of the liquid with nuclei. The breakaway of individual crystals was visible to the unaided eye; there were presumably other particles too small to be seen.

Similar effects were seen in vessels of diameters between 30 and 150 mm.

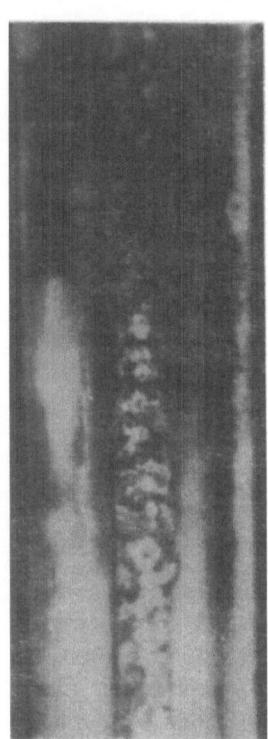

Fig. 36. Crystallization of benzophenone in the form of a thin film on the walls of a tube in response to insonation.

Although cavitation bubbles occurred in the melt (they were visible to the unaided eye), the crystallization occurred mainly at the sides and base of the vessel; the product was very inhomogeneous.

It was of interest to elucidate whether the nodes or antinodes are responsible for producing most of the nuclei. To establish this, the surface of the vibrator (an antinode) was coated with a thin layer of melted benzophenone; the result was that at some points there arose miniature fountains, but there was no appreciable effect on the crystallization rate, except at the sites of the fountains, where the layer of solidified material was somewhat thicker. The conditions at antinodes do not favor the production of fresh nuclei; experiments with other substances have given similar results [62].

Bagdasarov [57] examined the effects of wall vibration for the crystallization of melted p-bromaniline; tubes of diameters 5, 10, and 30 mm were used. The grain size was much reduced for the tubes 30 mm in diameter, but the walls were not found to have any effect for tubes 5 and 10 mm in diameter. The reason for this appears to be that the natural frequency of the 30-mm tube was close to the frequency of the ultrasound; the resonance accentuated the effect.

These studies have shown that vibration of the walls of the vessel can accelerate crystallization, as can cavitation and motion of gas bubbles; the wall effect is the most pronounced of these if the intensity is low. This interesting result should find some practical application.

CHAPTER IV

Effects of Various Agents on Nucleation in an Ultrasonic Field

1. Nucleation in Purified Materials

Kinetic studies have shown that an ultrasonic field affects the nucleation rate as well as the growth of the nuclei in a supercooled medium.

Nucleation in a supercooled medium may be spontaneous or induced under ordinary conditions; the latter requires the introduction of a seed, but the first occurs via random association of molecules. A group of molecules becomes the basis for the growth of a crystal if conditions are favorable. Until recently, it was not clear whether crystallization can occur in the complete absence of seeds of any kind or whether it is necessary to have present particles of the same substance, of an isomorphous substance, and so on. It was also uncertain whether crystallization could occur without seeds in response to sound, light, contact with solids, and so on.

Danilov [19] showed (for o-chloronitrobenzene) that insonation can accelerate the production of the first crystallization center if the melt is near the boundary of metastability.

I have used this compound, $C_6H_4NO_2Cl$, whose melting point is 31.5°; it is one of those substances that cannot be produced in vitreous form by any rate of cooling. Moreover, when purified it shows a sharp limit of metastability; it cannot be supercooled below 17°C.

In this respect, the compound resembles many metals.

The material was purified by repeated recrystallization and filtration (Schott filters No. 2 to No. 5). Of 14 specimens subjected to preliminary tests, I selected those that satisfied the requirements for spontaneous crystallization.

It is considered solely on the nature of the substance (is a characteristic constant). This rate should not be dependent on the time spent at a given supercooling; the absence of impurity activation and deactivation effects also indicate that the nuclei arise spontaneously.

My work on nucleation in this compound was guided by Gibbs's concepts on the formation of phases, as further developed by Frenkel' [63].

The theory of heterophase fluctuations and pretransition states indicates that nuclei of the new phase arise within the existing phase long before the two can be in the thermodynamic equilibrium; a system macroscopically homogeneous has a certain microinhomogeneity (has a certain density of nuclei), which varies little in time for given values of the thermodynamic parameters. Ultrasound may cause a redistribution of the micronuclei of the new phase, which would alter the probability of spontaneous production of crystals. The conditions at nodes and antinodes may have differing effects on the production of crystallization centers. Cavitation effects may influence the distribution, for example. These ideas have been tested by experiment.

The methods used with supercooled melts were as follows. The compound was heated to 100° on the water-bath and then was slowly cooled to 28, 25, 23, or 20°; on reaching the set temperature, the melt was irradiated for 10 sec. The intensity varied from run to run, but the frequency was always 700 kc. It was found that the field did not affect the production of the first crystallization center above 20°, although the intensity was varied within fairly wide limits; but the field caused spontaneous crystallization at 18-20°. The first center arose in different parts of the tube in repeat runs, and the delay varied from run to run; sometimes the first center arose as soon as the field was applied, and sometimes there was a delay of up to several seconds. The reason is that the waves alter the density of nuclei in the melt. Centers very seldom arose at high intensities (8 kV on crystal); sometimes it was necessary to apply the field 3 or 4 times. The conclusion is that ultrasound affects the production of the first crystallization center near the limit of metastability [64].

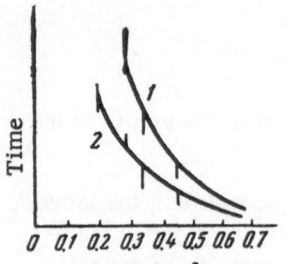

Fig. 37. Relation of time for appearance of first crystallization center to intensity and supercooling.

2. Nucleation at Impurities

Bagdasarov [57] has examined the effects of ultrasonic energy on nucleation at impurities for supercooled thymol and benzophenone. The impurities used were quartz sand and activated charcoal. The conclusion was that crystallization at the impurities does not occur in response to ultrasonics if the impurities have not previously been in contact with the crystalline material; but there may be an acceleration of nucleation if the impurities have been in contact with the solid. Bagdasarov found that the delay in the formation of the first center was reduced as the intensity was increased (frequency about 700 kc). Moreover, the time needed at a given intensity decreases as the supercooling increases, but the differences become less as the intensity increases and are very small at 0.65 W/cm^2 (p-bromaniline). Figure 37 shows the delay in the appearance of the first crystallization center as a function of intensity; curve 1 is for 35° and curve 2 for 47°. The reverse trend (increase in delay with intensity) is found for small degrees of supercooling (about 6°); this may be associated with heating consequent on absorption of the ultrasonics, for heating tends to deactivate impurities. In general, the ultrasound greatly reduces the width of the region of metastability.

★ 3. Nucleation in Amorphous Materials

The Tamman curve [65] (for the number of centers as a function of supercooling) indicates that the number at first rises but later falls almost to zero when the maximum has been passed. This means that a liquid not capable of spontaneous crystallization can remain in the amorphous (vitreous) state for a long time.

The amorphous state is usually produced by cooling the melt rapidly; the internal friction increases so greatly that the molecules are unable to form a crystalline lattice. The vitreous state is that of a supercooled liquid; thermodynamically speaking, it is unstable. The transition to the crystalline state can occur under suitable conditions, but only very slowly. For example, some ancient glasses show clear signs of crystallization; this indicates that glass contains a variety of atomic assemblies capable of acting as submicroscopic crystallization centers. The number, size, and composition of these are at present unknown.

A substantial advance in our understanding of the vitreous state occurred when it was appreciated that the structure of the material is as important as the composition. This means that effective methods of influencing the structure of glass are of practical and theoretical interest.

It is to be expected that ultrasound will affect the number and the growth rate of the crystallization centers [66].

The processes have been examined by reference to piperine, which can be converted to vitreous form by rapid cooling.

The piperine was heated to 140° and then was kept in a water-bath at 20° for 6 min, the temperature of the bath being kept constant to 0.1°. The rapid cooling converted the piperine completely to the vitreous form. The crystallization centers were caused to develop by transferring the specimen to a bath at 90° for 2 min, after which the number of centers could be counted.

This method was used to establish the effects of insonation; here the vitreous material was treated for 15 sec before development.

Irradiation at a frequency of 700 kc caused far more centers to appear, but there was no increase in the number of centers when the frequency was 30 kc.

CHAPTER V

Growth and Dissolution of Monocrystals

1. Effects of High-Frequency Ultrasound on Crystallization and Dissolution

Ultrasound affects crystallization by producing many fresh nuclei from any seed that may be present in the supercooled medium. The number of nuclei can be controlled within certain limits, and this provides a means of altering the polycrystalline structure of the product.

The growth of a monocrystal in an ultrasonic field has many interesting aspects. A growing crystal receives molecules or molecular assemblies only from the medium in contact with it; the principal factors here are diffusion and concentration currents (solutions) or self-diffusion and thermal conductivity (melts). A current topic of discussion is whether ultrasound can influence molecular exchange in general and diffusion in particular; a decision on this question would enable us to elucidate many known effects of ultrasonics. Moreover, the changes in the surfaces of solids produced by ultrasonics provide a means of examining the effects of plastic deformation on the growth of crystal faces.

The structure of a face is made up of a variety of elementary and macroscopic steps, together with other features. The height distribution of the steps is one of the main characteristics of the surface; it is very much dependent on the conditions (supersaturation, deformation, presence of impurities, and so on), and it influences the properties of the final monocrystal.

New active centers of growth can arise at the points of application of the forces in a deformed crystal.

Measurements have been made [67] of the effects of ultrasonic energy on the linear growth rate of the octahedron faces of potash alum monocrystals; the rate was measured on a crystal in the field and on another under identical conditions apart from the field. The source was a quartz plate driven at 2 Mc, and the maximum intensity was 0.2 W/cm^2.

Two identical crystallizers were used; each was a glass vessel (capacity 300 cm^3) having double walls, the outer jacket containing circulating water to control the temperature. The same water passed through both jackets in series. The positions of the growth fronts were measured to 0.01 mm with cathetometers. The temperature was kept constant to 0.1°. The saturation point of the solution was determined by the concentration–current method (the Toepler effect was used). The linear growth rates of the two crystals were the same in the absence of the field.

The solution at a temperature 7-8° above the saturation point was poured into the preheated crystallizers, which were then rapidly closed by covers providing for mounting of crystal holder and thermometer. Each crystallizer was fitted up with a weighed seed; the ultrasound was turned on in one crystallizer, and the cathetometers were adjusted to view the faces. The linear displacements of the faces were measured at intervals of 15-30 min at first, and then once an hour for 13-15 h; the crystals were then withdrawn, dried, and weighed, to evaluate the rate of increase in mass.

The same procedure is, of course, applicable to the dissolution of crystals and was so used.

It was found that the ultrasound accelerated dissolution when the temperature of the solution was above the saturation point; the mean linear rate of displacement of the faces was roughly doubled by the field.

Conversely, the growth rate in the field was higher when the solution was supersaturated; the ultrasound increased the mean linear growth rate by a factor of 2-3. The gain in weight in the ultrasonic field was 50-60 mg in 10-15 hr. Figure 38 illustrates the general behavior. The field accelerates the growth of faces turned towards the source and also of other faces (see below).

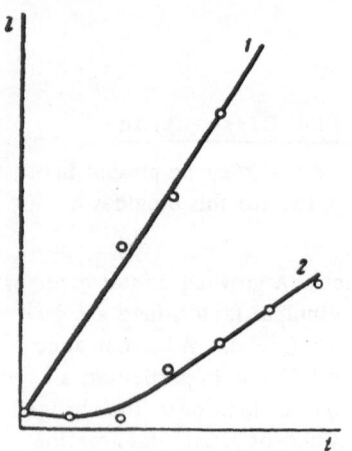

Fig. 38. Effect of ultrasound on the growth of alum monocrystals: 1) Face growth in field; 2) face growth without field.

The ultrasound has this pronounced effect only when the degree of supersaturation is low and the field is relatively weak; the usual dispersal of the crystal occurs at high intensities. ✓

Kavalyunaite has examined the effects of intensity and frequency on the growth and dissolution of alum monocrystals.

Two crystals (weights 27 and 27.2 mg) were placed in two crystallizers containing slightly warm solution; the temperature was then reduced slightly to initiate growth. The beam (1.43 Mc) was directed parallel to one face of the crystal. Figure 39 shows the linear displacement and linear rate, the first being the actual measurements and the second obtained by differencing the results. The growth rate of the octahedron faces is not affected by the field while this rate is less than 0.023 mm/hr, but it is reduced at higher rates, the effect increasing with the rate. The gain in weight Δm was 344 mg for the crystal in the field and 288.3 mg for the other crystal, so the ultrasound increased the gain in weight by a factor 1.2 relative to ordinary conditions.

Another series was performed at 2 Mc with solutions having various degrees of supersaturation and with seeds having equally developed faces (weights of pairs: 20 and 20.8 mg; 20.6 and 20.8 mg; 18 and 18 mg). The faces were observed after the seeds had regenerated. The temperature was reduced from 25 to 24° in the first run, from 25.6 to 23.2° in the second, and from 25.5 to 20.6° in the third. Figure 40 shows the results. The linear rate for the control crystal in the first run was 0.021 mm/hr, that for the irradiated crystal being 1.8 times larger; the second run gave a control rate of 0.03 mm/hr and no effect from the field, while the third gave a control rate of 0.09 mm/hr and a rate smaller by a factor 1.9 for the irradiated crystal. The ultrasound produced a larger increase in weight in every case (first run, by a factor 1.8; second, 1.4; third, 1.3). The field thus loses its effect on the linear rate for this particular face at a certain supersaturation, but it retains a reduced effect on the other faces, as the greater increase in mass indicates.

The effects of intensity were examined by means of seeds (weight about 23 mg) used with solutions of saturation temperature 25.5°; the ultrasound was introduced when the solution had reached 25°, which caused appreciable growth of the face. The intensity was varied in steps between 0.006 and 0.25 W/cm². Figure 41 shows the results.

The broken lines on the curves for v indicate the times when the intensity and supersaturation were changed.

Low intensities (mW/cm²) accelerated the growth (part 1-2 on curve) if the control rate was less than 0.02 mm/hr, but the effect became zero at a control rate of 0.027 mm/hr (point 2). The growth rate was again raised when the intensity was increased (part 3-4), but further increase in intensity caused a fall in the rate (part 5-6), and several crystals appeared at the bottom of the crystallizer.

Another series was performed at 3 Mc with seeds of initial weights 27.6 and 27.8 mg (saturation temperature 25.5°). The results were much the same, but the maximum growth rate of the octahedron face was somewhat larger than that for 2 Mc.

The size of the seed also influences the result. In one case, the seed weighed 13.4 mg, while other seeds had weights of 50.1 and 50.9 mg and had well-developed faces. In both cases the saturation temperature of the solution was 26.5°. The crystals were grown for the first forty hours with a temperature reduction from 25.5 to 23.7°. The ultrasound accelerated face growth in a small seed if the control rate was below 0.038 mm/hr; 0.04 mm/hr is the limit at which the ultrasound has no effect. On the other hand, the ultrasound actually retarded the growth of large seeds when the control rate was 0.037 mm/hr.

In the first case, the irradiated crystal gained 1.5 times as much weight as the control; in the second, the rate was only 1.06 for the same conditions.

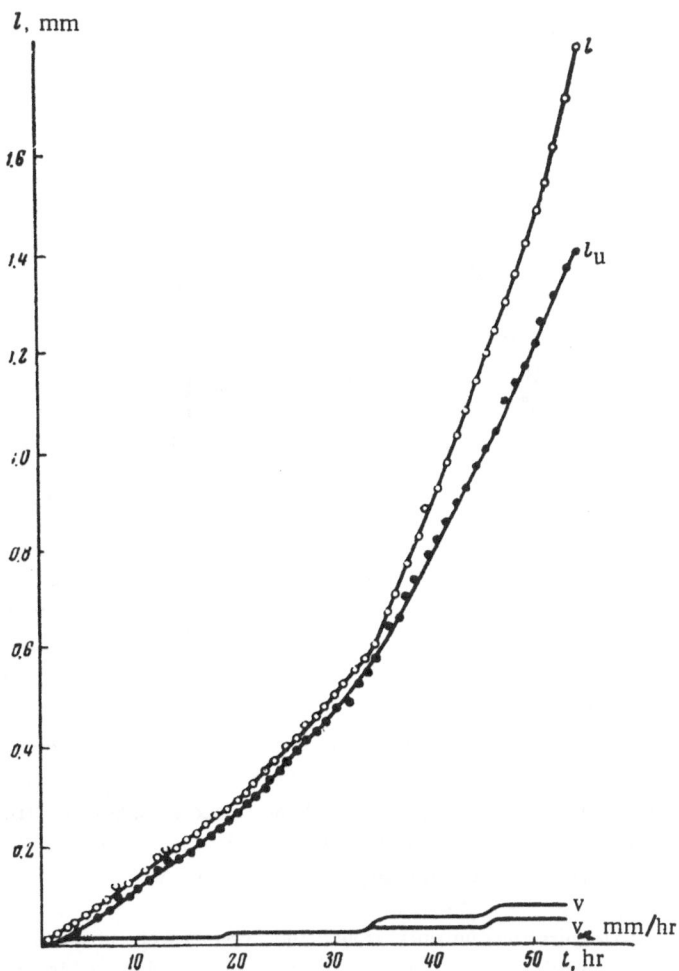

Fig. 39. Linear displacement and linear growth rate of faces as functions of time: l and v) No ultrasonic field; l_u and v_u) ultrasonic field applied.

2. Effects of Low-Frequency Ultrasound on the Growth of Monocrystals of Potash Alum

Low frequencies do not have the same effects as high ones; the size of the seed was about 3 times the wavelength for the high frequencies but was less than 1/10 of the wavelength of the low frequencies. Moreover, cavitation occurs more readily at low frequencies, and this gives rise to new effects.

Our experiments at low frequencies were done with intensities such that cavitation could not occur (0.2 W/cm^2 at 30 kc); the crystals were placed at the nodes and antinodes of standing waves, the seeds being of masses 22 and 22.3 mg. The initial growth rates (before the field was applied) were 0.06 mm/hr in both cases; the temperature during the run was reduced from 23 to 22.2°, the saturation temperature being 26°. Figure 42 shows the results; the Δm for the crystal at the node was 250 mg, that for the control crystal being 263.7 mg (ratio 1.04). The crystal placed at the node was somewhat wider in directions normal to the direction of propagation.

The crystals used in the antinode test had initial weights of 21.2 and 21.4 mg; here the ultrasound increased the linear growth rate. The ratio of the increases in mass (irradiated to control) was 1.05; here the irradiated crystal was elongated parallel to the direction of propagation. Tangential growth predominates at nodes; normal

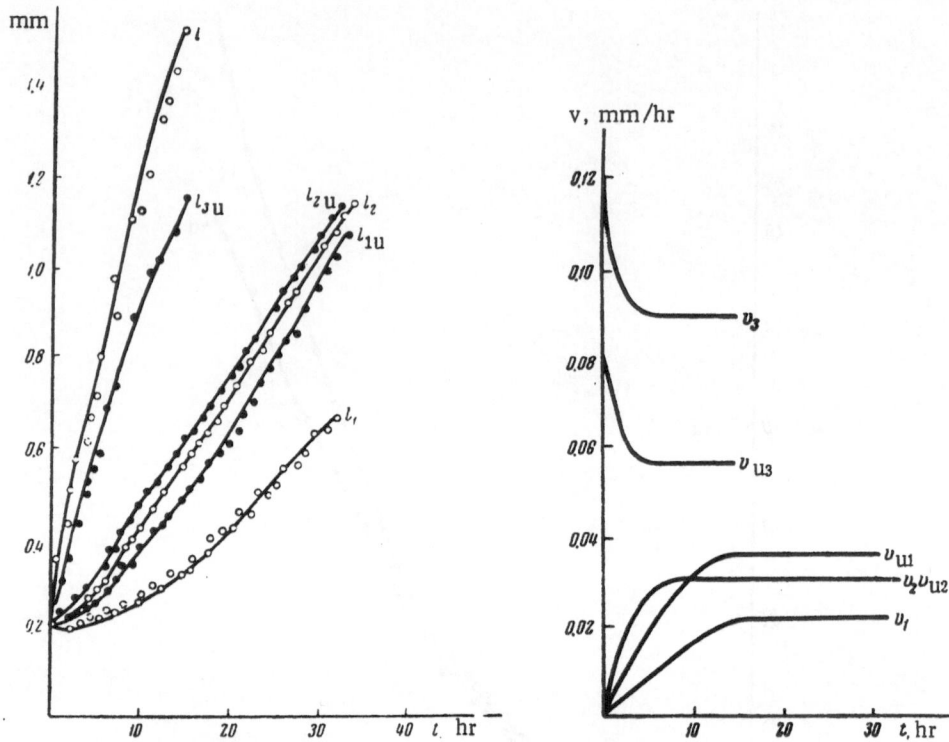

Fig. 40. Linear displacement of faces (left) and face growth rates (right) as functions of time for potash alum. The number subscripts denote the corresponding runs; subscript u denotes values for crystals exposed to ultrasound.

growth, at antinodes. The concentration currents around the crystals were nearly straight and were almost unaffected by the field at nodes, but the currents were deflected at antinodes, as Fig. 43 shows. It is clear from Fig. 43a that the crystal lies at a node, and from Fig. 43b that it lies at an antinode.

3. Avalanche Crystallization of Potash Alum from Solution as Influenced by Ultrasound

Studies of effects of insonation (30 kc) on the crystallization kinetics of solutions have enabled us to detect some effects that throw some light on the growth mechanisms of monocrystals.

A solution 6-8° below its saturation temperature gives rise almost at once to numerous fresh nuclei when the ultrasound is switched on if a monocrystal seed is present and if the intensity is sufficiently high [67]; the crystals soon fill the entire volume (Fig. 44). The same occurs if the frequency is 700 kc. This avalanche nucleation indicates that the solution is in an unstable state; the ultrasound gives rise to so many nuclei that the solution becomes opaque. A certain time elapses before crystals large enough to see are observed sinking slowly. The supernatant remaining after the crystals have settled out has a saturation temperature equal to the room temperature or 1-2° above it. The crystals take the form of needles and plates. The effect is very much dependent on the intensity, the concentration, and the position of the seed. Intensities below 0.2 W/cm² produce the effect only if the supersaturation is high (supercooling of 6-8°); low degrees of supercooling (1°) suffice above 0.2 W/cm². The cloud appears more readily and more rapidly increases in size as the intensity and supersatura-

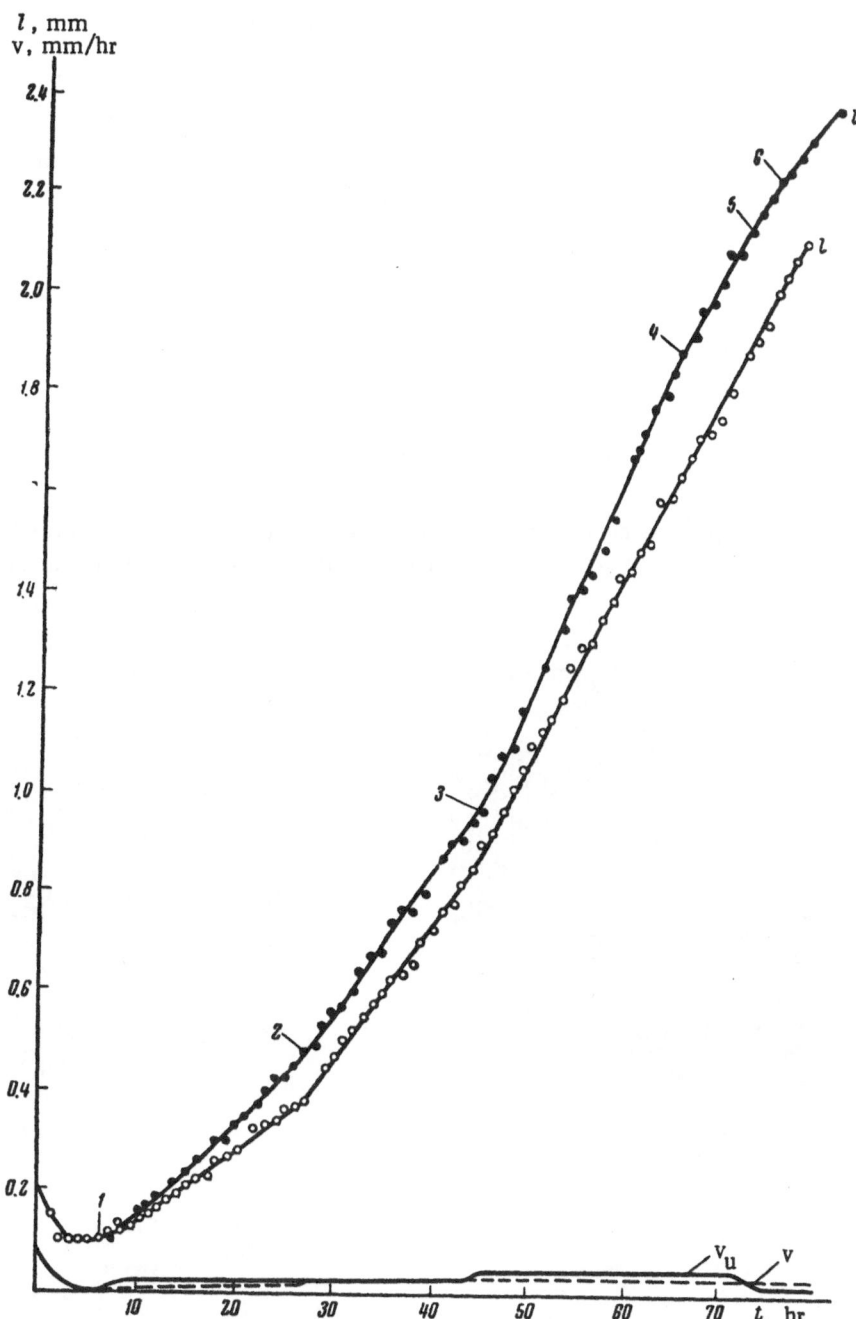

Fig. 41. Linear displacement and linear growth rate of faces as functions of time: l and v) No ultrasonic field; l_u and v_u) ultrasonic field applied.

tion increase; the same occurs if the seed is moved from a node to an antinode. The avalanche effect is maximal if there is a stable standing-wave pattern; this can be ensured by adjusting the depth of solution.

Some details are of interest. A patch of small crystals appears at the adjacent antinode when the seed is at an antinode if the supersaturation is appropriate; the patch receives further small crystals from the surrounding medium but loses its larger ones. The patch falls to the bottom if the intensity is increased. A large crystal inserted with a face normal to the waves will bind firmly some of the small crystals that reach it.

In such cases, most of the attached crystals form parallel overgrowths on the main crystal (Fig. 45).

Shakol'skaya and Shubnikov [68] have reported a similar effect, but in our case the crystals were much smaller, for they had no time to grow. Shubnikov [4] considers that these experiments are important to the theory

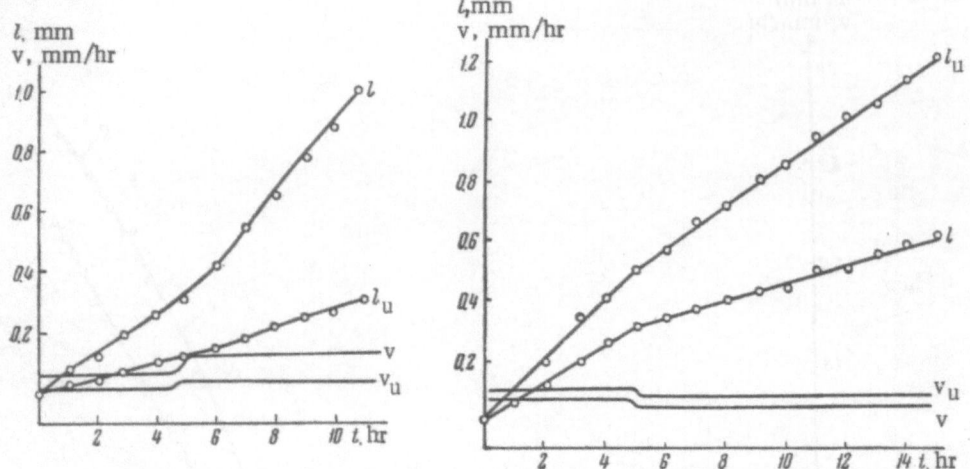

Fig. 42. Linear displacement of octahedron faces and face growth rates as functions of time; subscript u denotes values for crystals exposed to ultrasound. Crystal at node, left; crystal at antinode, right.

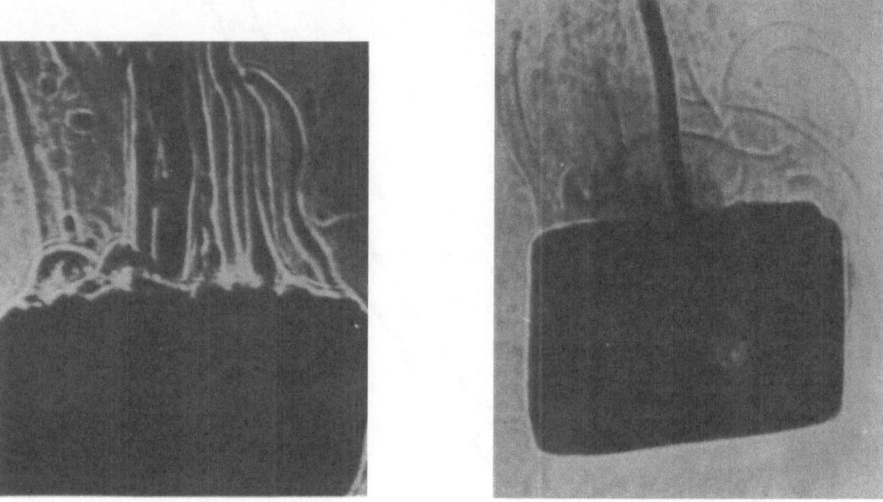

Fig. 43. Crystallization currents around a crystal: a) At a node; b) at an antinode.

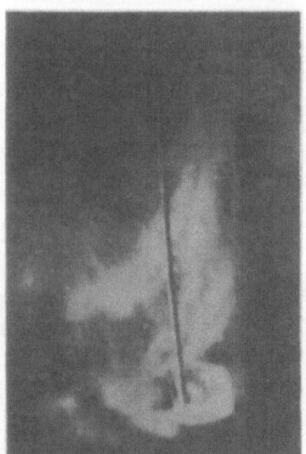

Fig. 44. Production of small crystals by ultrasound.

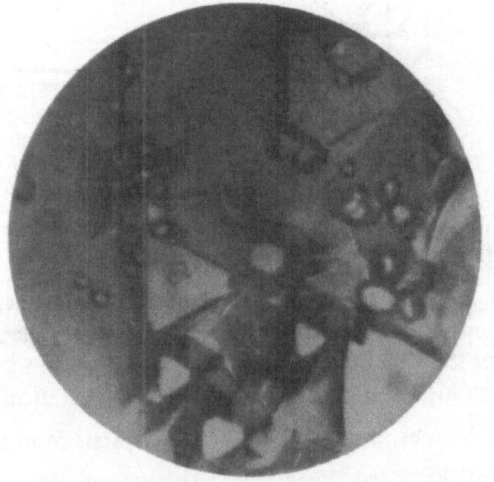

Fig. 45. Parallel intergrowths of alum crystals.

46

of crystal growth, for they show that the lattice of a real crystal [82] is not built up from isolated atoms and molecules alone; visible and invisible assemblies may join on as parallel overgrowths, which afterwards cannot be distinguished.

4. Properties of Metal Monocrystals Grown in an Ultrasonic Field

Bridgman's method has been used to grow monocrystals of zinc in an ultrasonic field at 800 kc [69] (Fig. 46). The apparatus consisted of a turbular furnace, a quartz holder, and an ultrasonic source.

A zinc monocrystal previously grown in a sealed glass tube 3-4 mm in diameter was attached to the top of the quartz holder. The top of the crystal entered the oven, the lower part acting as the waveguide. The drive mechanism and ultrasonic generator were turned on when the top of the crystal had melted; the crystal was driven at 37 mm/hr to cause it to melt and then crystallize again while acted on by the ultrasound. The maximum furnace temperature was 520°. The experiment was continued until the crystal had left the furnace completely. The crystal was then tested for strength in comparison with a relaxometer developed in the Laboratory of the Mechanical Properties of Crystals, Institute of Crystallography, Academy of Sciences of the USSR.

Figure 47 (curve a) represents the compression of a crystal grown with the ultrasound applied; curve b is the same, but for a crystal grown without the ultrasound. The treatment clearly increases the strength considerably; the elastic limit is increased by nearly a factor 6. A crystal grown without the field and then irradiated was also used (curve c of Fig. 47); here the elastic limit was nearly doubled. Crystals treated in these two ways were annealed at 350° before test in order to establish the cause of this effect. Figure 48 shows the results, curve a being for the crystal grown in the field and curve b for the crystal irradiated after growth.

The hardening effect is reduced only in the case of the crystal grown and then irradiated.

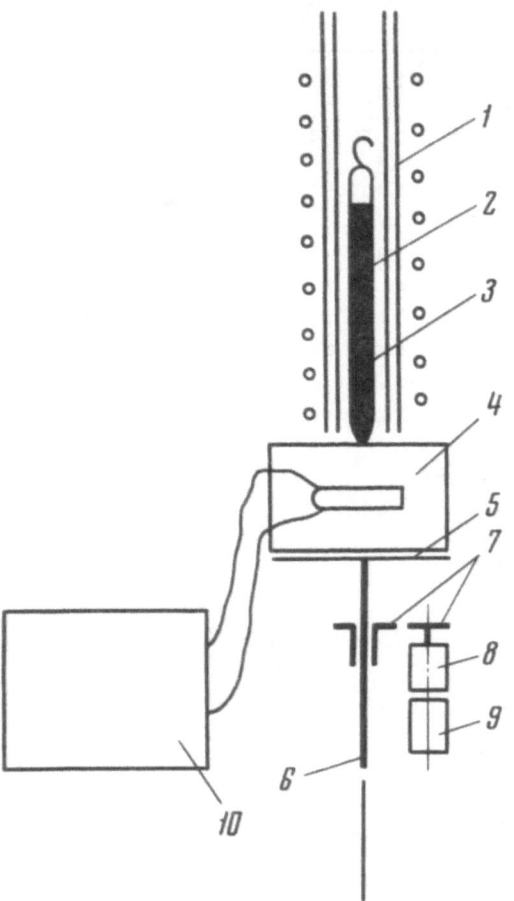

Fig. 46. Apparatus for growing monocrystals in an ultrasonic field: 1) Resistance furnace; 2) glass tube; 3) substance; 4) quartz holder; 5) support; 6) screw; 7) gears; 8) reduction gear; 9) motor; 10) ultrasonic generator.

5. Dissolution of Crystals in an Ultrasonic Field

Dissolution is one aspect of phase transitions generally; ultrasound accelerates diffusions and heterogeneous processes, so it is natural to expect that it influences dissolution.

Schmid and Roll [33] have shown that ultrasound (9 kc) increases the rate of solution of iron in molten zinc very considerably. Zinc contained in an iron crucible at 500°C very soon takes up 1% of iron when it is irradiated. Table 2 gives the iron content of the zinc as a function of insonation time.

Pfefferkorn [70] observed that ultrasound causes etching agents to attack the surface of the crystal more evenly. I have followed up this observation with tests on crystals of sugar, copper sulfate, thiosulfate, and thymol at 500 kc and 2 W/cm^2 (Table 3).

In all cases, the ultrasound accelerated the process greatly.

Table 2

Insonation time, min	% iron in zinc
5	0.2
10	0.6
20	0.8
30	1.0

Table 3

Substance	Mass, mg	Time to complete solution, sec		Temp., °C
		ultrasound off	ultrasound on	
Sugar	470	600	60	70
Copper sulfate	250	1200	60	24
Thiosulfate	139	23	7.8	90
Thymol	280	55.8	19.6	90

Ultrasound also affects the rate of attack of mineral acids on steel [71]. Thin rectangular strips of steel were cleaned with emery paper and were washed with water, alcohol, and ether; then they were weighed. The plates were placed in 1 N sulfuric acid and were irradiated for times of 5 to 25 min at 100 kc and 4 W/cm². Figure 49 shows the results.

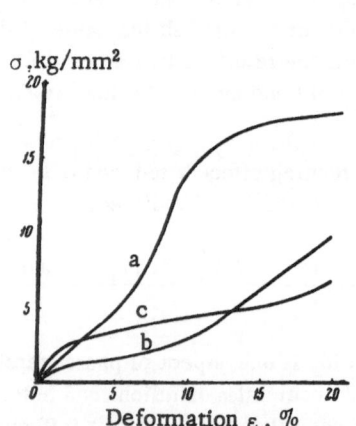

Fig. 47. Compression curves for crystals grown with and without the application of ultrasound.

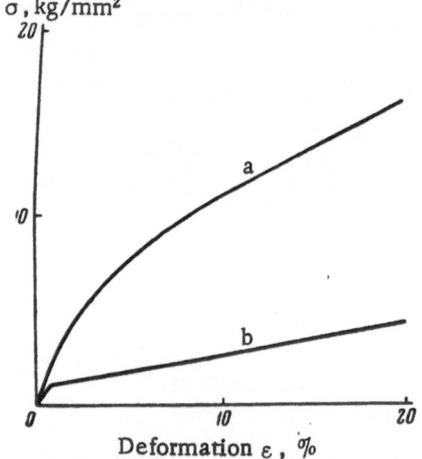

Fig. 48. Compression curves for crystals grown with the application of ultrasound but recorded after annealing.

These curves show that the amount dissolved is proportional to the time of contact, and also that the ultrasound accelerates the process considerably (curve a). For instance, the amount dissolving in 25 min in the field is about 3.5 times that dissolving without the field (curve b).

The rate of attack is also strongly dependent on the intensity. Figure 50 shows the rate as a function of intensity for steel; low intensities clearly have little effect on the rate.

There is a rapid rise above a certain intensity, an upper limit then being approached.

Tsyganova and Lebedev [72] examined the corrosion of aluminum in an ultrasonic field. Polished plates of area 18 to 25 cm² and 0.95 mm thick were immersed in 19.3% HCl; two plates were irradiated (20 and 700 kc) and a third acted as control. The irradiation time was 100 sec at 20°. Figure 51 shows the results. The rate of attack was largest for the control plate and least at 700 kc.

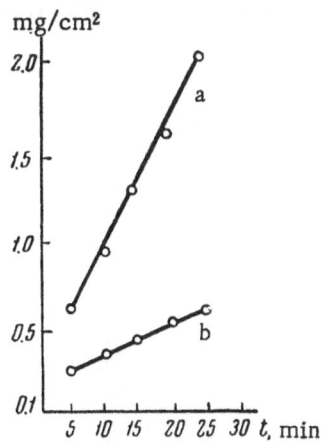

Fig. 49. Attack of mineral acid on steel: a) Ultrasound on; b) ultrasound off.

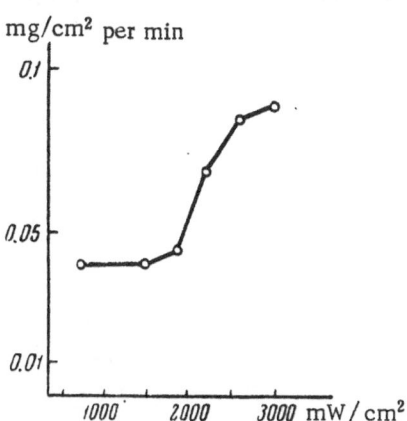

Fig. 50. Effects of ultrasonic frequency on the solution of steel in acid.

Yamaguchi [73] reported similar results for magnesium specimens in 3% NaCl exposed to 1 Mc (3.5 W/cm²); the irradiated material showed less surface damage than the control. The results were explained in terms of a closed protective layer.

Meiswinkel [74] used frequencies of 3 to 30 kc at about 2 W/cm² in studies of the etching of thin metal plates; he considered that the ultrasonics provided a 30% economy in the use of acid, with halving of the etching time.

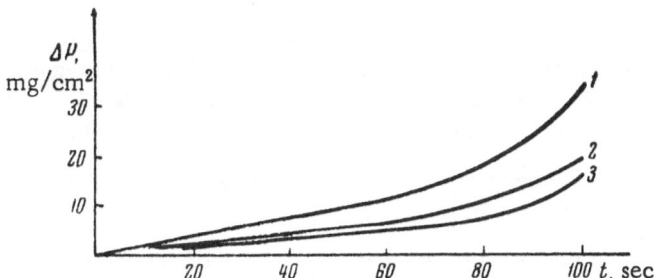

Fig. 51. Dissolution of aluminum as a function of time and ultrasonic frequency, kc: 1) 0; 2) 20; 3) 700.

Bagdasarov observed the dissolution of a monocrystal of rock salt in water at room temperature at frequencies of 22, 410, and 717 kc; the rate was higher at the lower frequencies, that at 22 kc being almost 7 times that at 410 kc. The effect is associated with dispersal, which is much more strongly developed at low frequencies.

6. Production of Etch Figures in the Presence of Ultrasound

It has been found that ultrasound affects the initial stage of dissolution of a monocrystal, namely formation of etch figures. The etch-figure pattern is a result of the variation in solubility with direction; it often enables one to deduce the symmetry of the crystal. There are many papers on etch figures [75], but the precise causes of the greater reactivity (i.e., etch pits) at some points are not known. It is now generally considered that they are associated with sites of defects (including dislocations) in the lattice.

Etch figures are often produced by the use of unsaturated solutions of the material, but this method sometimes fails to yield the figures, in which case special reagents are used; but these often alter the shape of the figures greatly.

Fig. 52. Etch pits on an octahedron face of a crystal of potash alum.

Ultrasound has been used to obtain etch figures [76]; the crystal is placed in a saturated solution and is irradiated for 2-3 min at low intensity (e.g., 40 kc at about 0.3 W/cm^2 from a magnetostriction source). Etch figures on the octahedron faces of potash alum have been produced in this way (Fig. 52); so have ones on cleavage planes parallel to the pinacoid in benzophenone, and also on irregular crystals of potassium dichromate.

Etch figures can be produced at any stage in dissolution, on account of increased rate of attack at defective sites, as Bagdasarov and Khaimov-Mal'kov have shown [77].

Further work has shown that etch figures may sometimes be better produced at higher frequencies, up to 2 Mc.

For example, Kameneva [78] has used potash alum monocrystals at 2 Mc with intensities of 0.017 to 0.25 W/cm^2. The ultrasound was found to affect the shape of the etch pits on the octahedron, cube, and rhombododecahedron faces; it also affected the development of the figures in pure water and in solutions of alum. Figure 53 shows etch pits produced on the octahedron and rhombododecahedron faces by distilled water at 18° in 30 sec. The rate of attack increases with the intensity. The intensity also affects the shape of the pits as produced in water (Fig. 54) and in solutions (Fig. 55).

The rate of attack falls as the solution becomes more concentrated; above a concentration of about 8% (3% away from saturation) the ultrasound has no effect on the rate, but the shape of the pits is affected (Fig. 56).

The rate of attack is then independent of the intensity, but the shape of the figures is still affected.

The state of the surface affects the distribution of the pits. For example, a scratch on the octahedron face of an alum crystal has many pits around it (Fig. 57). The orientation of these is the same near the scratch

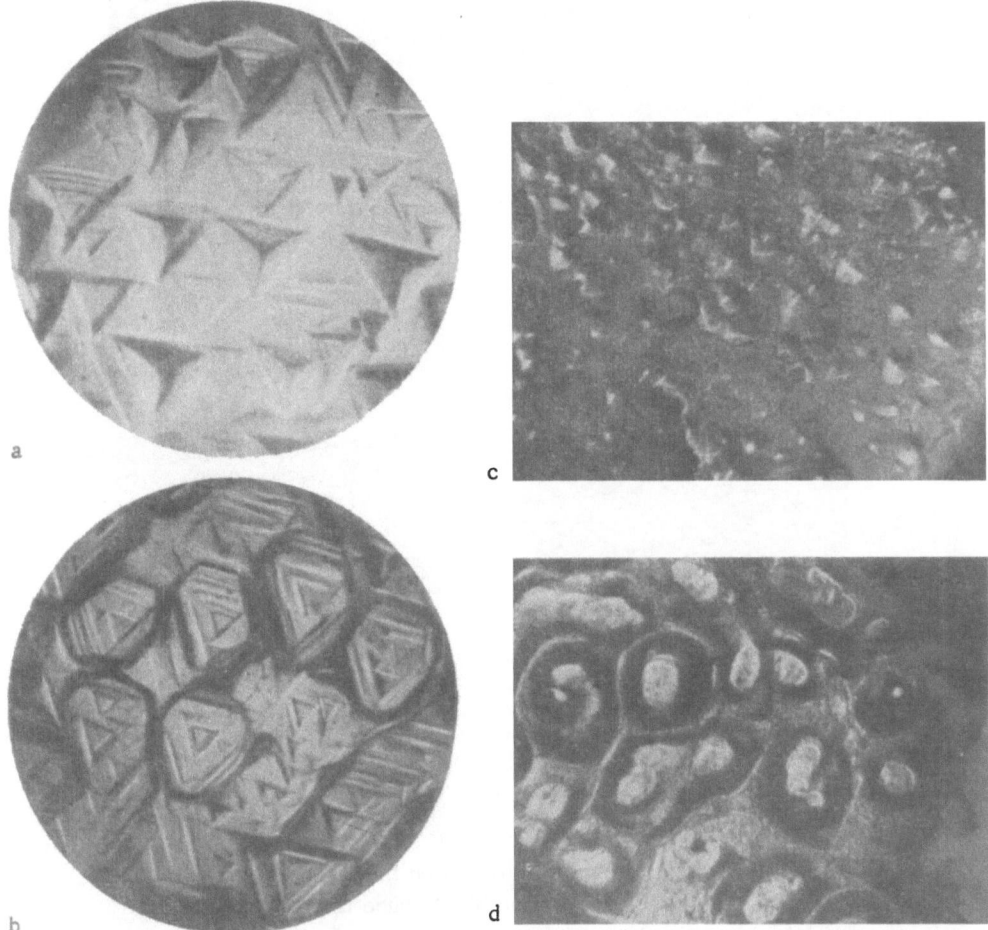

Fig. 53. Effects of ultrasound on the etch figures on potash alum: a) Octahedron face, no ultrasound; b) ultrasound applied; c) rhombododecahedron face, no ultrasound; d) ultrasound applied.

and far away. The pits near damage sites develop later than those far from sites of damage when the intensity is 0.049 W/cm^2; the same pattern is obtained without the ultrasound, but it develops more slowly.

Bagdasarov [79] has shown that allowance must be made for the ratio of wavelength to crystal size in any discussion of processes of growth and dissolution for monocrystals; the direction of the face relative to the direction of propagation is also important.

Faces parallel to the beam, or inclined to it, show regular series of ridges and hollows, whose spacing is governed by the angle between face and beam. This spacing between adjacent ridges or hollows is half the wavelength if the face is parallel to the beam.

Figure 58 shows the surface of a potash alum crystal exposed to standing waves at 2.5 Mc; the crystal was very much larger than the wavelength. The currents resulting from the attack are clearly visible below the crystal; these are split up into layers in accordance with the wavelength.

The ultrasound also affects the form of the diffusion layer near the surface; Fig. 59 shows the effect after the field has been applied. The concentration remains practically constant along the surface in the absence of the field, but it shows a periodic structure when the field is present. This structure is seen as ridges separated by equal distances, and it corresponds to the pressure distribution in the wave.

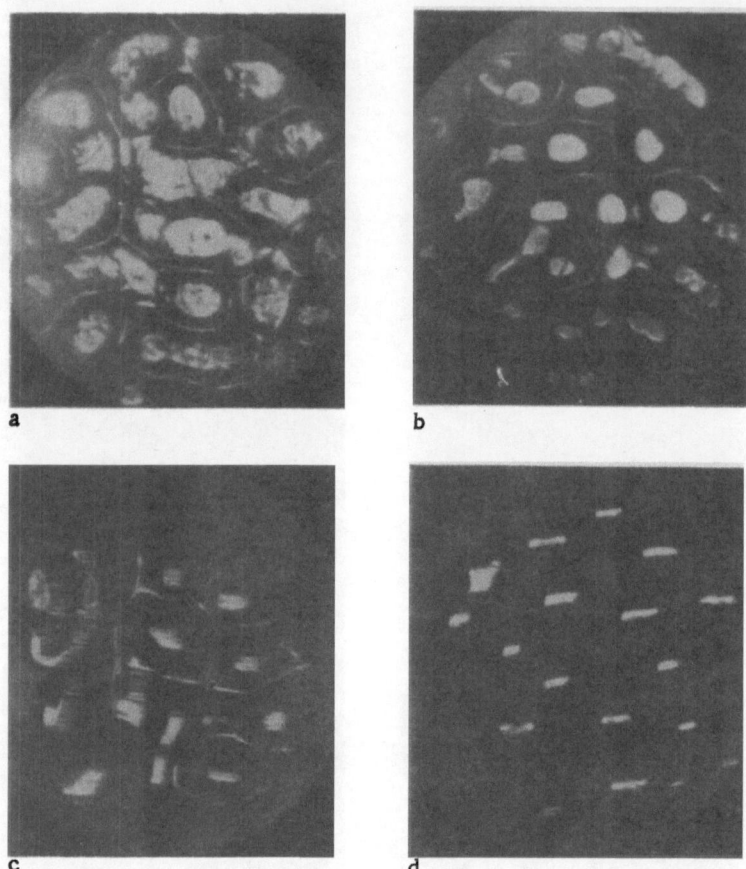

Fig. 54. Effects of ultrasonic intensity on the shape of etch pits produced by the pure solvent: a and b) On cube faces (0.017 and 0.25 W/cm², respectively); c and d) on rhombododecahedron faces (0.017 and 0.25 W/cm², respectively).

Bogdasarov and Khaimov-Mal'kov have made some novel observations on the nature of the etch pits produced in ultrasonic fields [77].

The figures are produced at any stage in the dissolution, which indicates that the ultrasound produces microcracks on the surface, these then acting as centers for etch pits.

An effect described by A. F. Ioffe has been used to demonstrate this. The low ultimate strength in rupture for rock salt crystals in air is caused by critical surface cracks, which are eliminated when the crystal is immersed in water, because the surface layer is removed.

This effect was used in rupture tests on crystals of NaCl cut along (100), which were tested in air and also in water with or without ultrasound applied (frequencies 717 and 22 kc). Table 4 gives results for tests made at 0.2 W/cm².

Table 5 compares the mean values with ones from the literature.

There was no appreciable change in the strength when the temperature of the water was raised to 80°.

These results show that the strength in water was reduced to the strength in air when the 22 kc was used; this is a result of the critical microcracks produced by cavitation at the surface, for the use of 717 kc, which causes less cavitation at this intensity, caused little deviation from the strength in water.

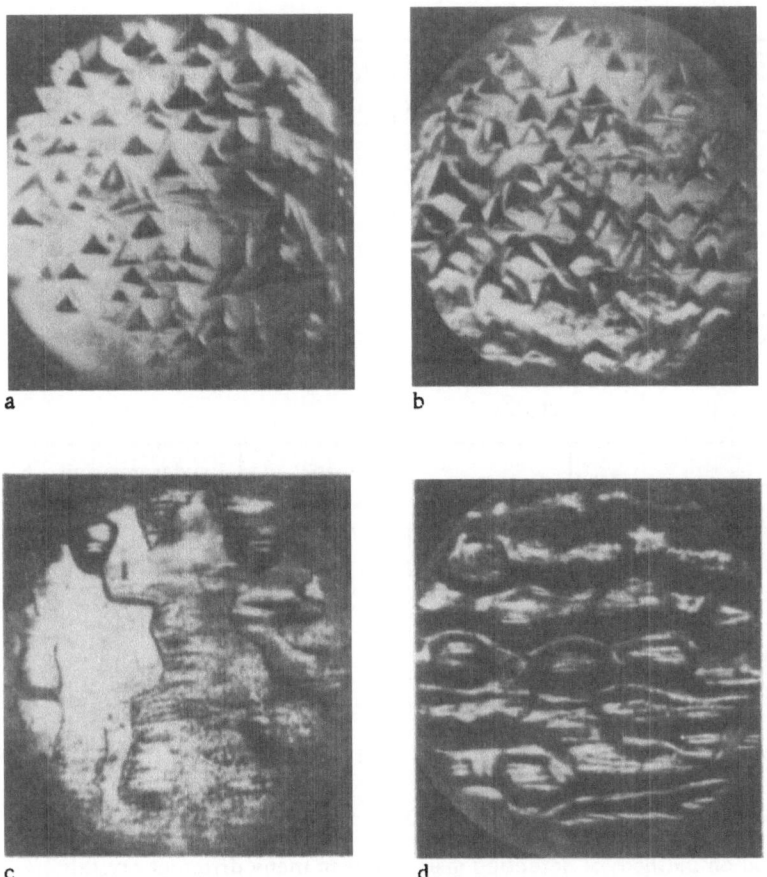

Fig. 55. Effects of ultrasonic intensity on the shape of etch pits produced by solutions: a and b) On cube faces (0.017 and 0.25 W/cm^2, respectively); c and d) on rhombododecahedron faces (0.017 and 0.25 W/cm^2, respectively).

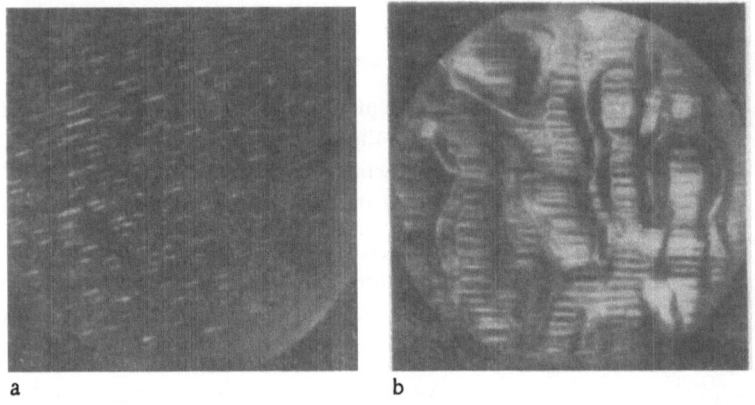

Fig. 56. Shape of etch pits under conditions such that the ultrasound does not affect the rate of attach (rhombododecahedron face); a) No ultrasound; b) intensity 0.017 W/cm^2.

Table 4

Strength dry, g/mm²	Strength in still water, g/mm²	Strength in running water, g/mm²	Strength (g/mm²) in water acted on by ultrasound at	
			717 kc	22 kc
398	2430	4960	3170	450
298	2130	5900	2980	200
510	2160	3850	4650	476
360	2500	3410	5370	150
430	2660	6180	–	–

Table 5

Mean strength, g/mm²	Ioffe	Klassen-Neklyudova	Bagdasarov and Khaimov-Mal'kov
Dry salt	400-500	294	399
Still water	–	–	2380
Running water	4000-5000	2486	4860
717 kc applied	–	–	4040
22 kc applied	–	–	319

Ultrasound thus tends to produce microcracks as a result of cavitation, and these act as nuclei for etch pits.

7. Detection of Dislocations in Crystals with Ultrasonics

Many papers have appeared on methods of detecting dislocation in many different crystals.

Selective etching is frequently employed; the mechanism of formation of the etch pits is related to the defects, and especially with the emergence of dislocations on the surface. In this way, Gilman and Johnston [80] have observed the migration of individual dislocations in LiF, for example, but they remarked that tedious trials are needed to establish the conditions under which the method is effective.

Mechanical vibrations enable one to detect active areas on the surface more rapidly.

Ultrasonics can assist in the production of etch figures [81]; the ultrasound accelerates the development of the pits, whose shape and position correspond precisely with those of the pits produced by the etching agent alone.

Our tests were done with LiF monocrystals, which do not produce etch pits when they are immersed in water for a short time under ordinary conditions. Insonation in water at 25 kc and about 0.5 W/cm² was used to reveal the pits.

The method of preparing and testing the crystals was very simple. Pieces 1.05 x 0.2 cm were cleaved along cleavage planes from a monocrystal; a freshly cleaved specimen was fixed at the end of a wooden rod and was placed at an antinode in the standing wave in the water.

Fig. 57. Etch pits on a plastically deformed alum crystal.

The (100) face was examined under a polarizing microscope after exposure; insonation for 15-20 min produced chains of pits of pyramidal form (Fig. 60). More prolonged insonation produced deeper pits, but the number did not alter greatly. Moreover, the two matting faces produced by cleavage from one specimen showed identical

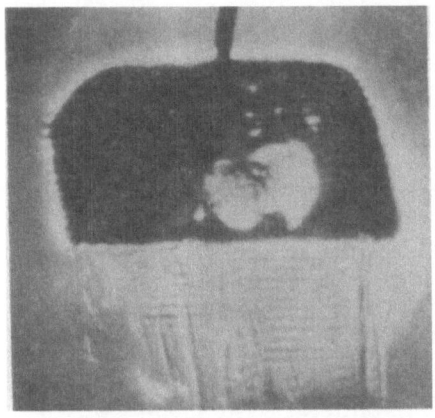

Fig. 58. Diffusion layer around a dissolving alum crystal in an ultrasonic field.

Fig. 59. Diffusion layer around an alum crystal growing in an ultrasonic field.

patterns after insonation. Further tests were done on large LiF crystals, which were repeatedly cleaved into thinner and thinner plates, all of which were irradiated at 0.5 W/cm². Even the thinnest plates gave the pyramidal pits. The effect was reproducible for cleavage on other crystallographic planes also.

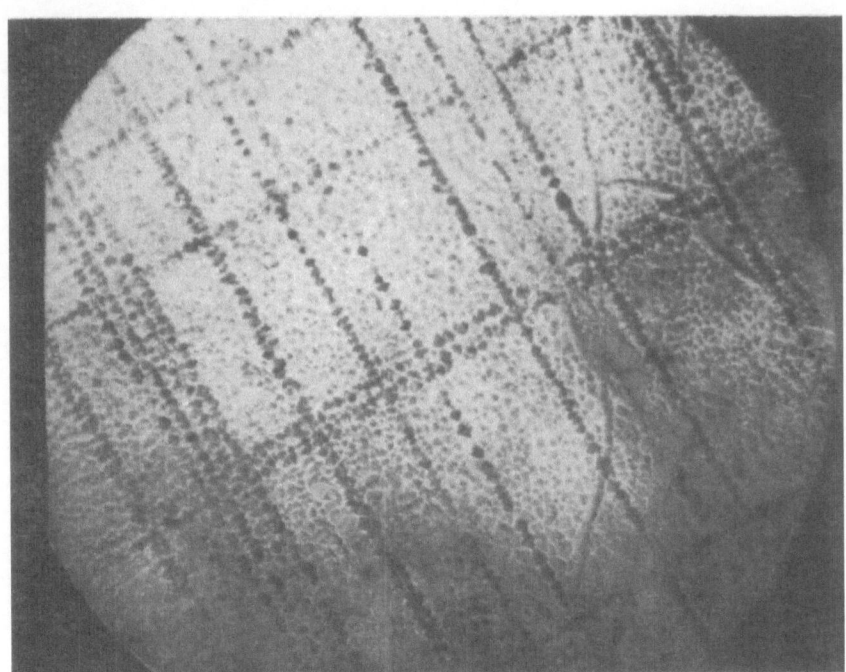

Fig. 60. Chains of etch pits on the surface of a crystal of LiF, × 270.

Oval pits appear in response to very prolonged insonation; the depth and diameter vary, and the number is larger near the edges. Figure 61 shows a typical pit of this kind, which may be 2-3 mm deep. Chains of smaller

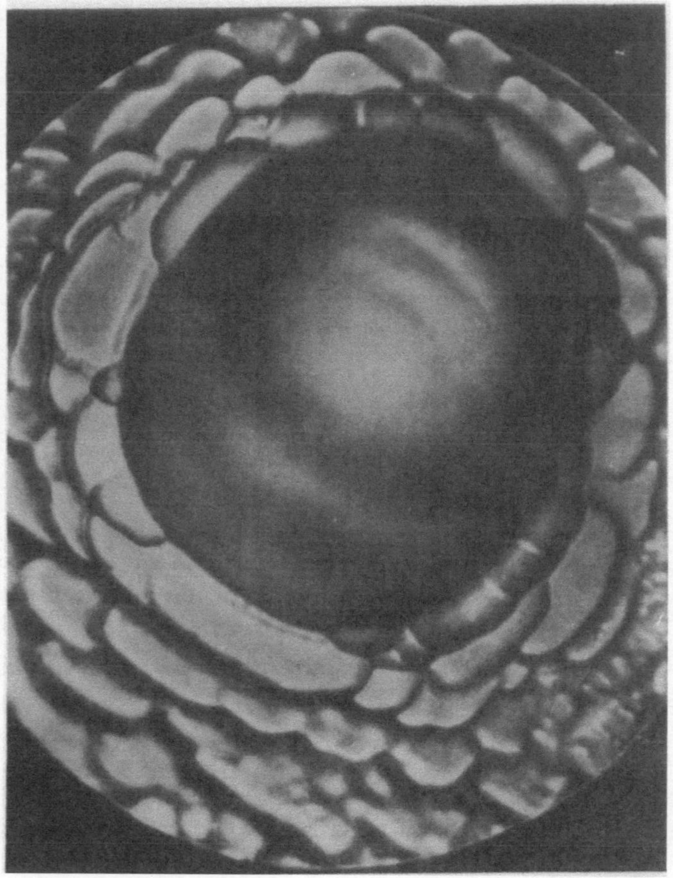

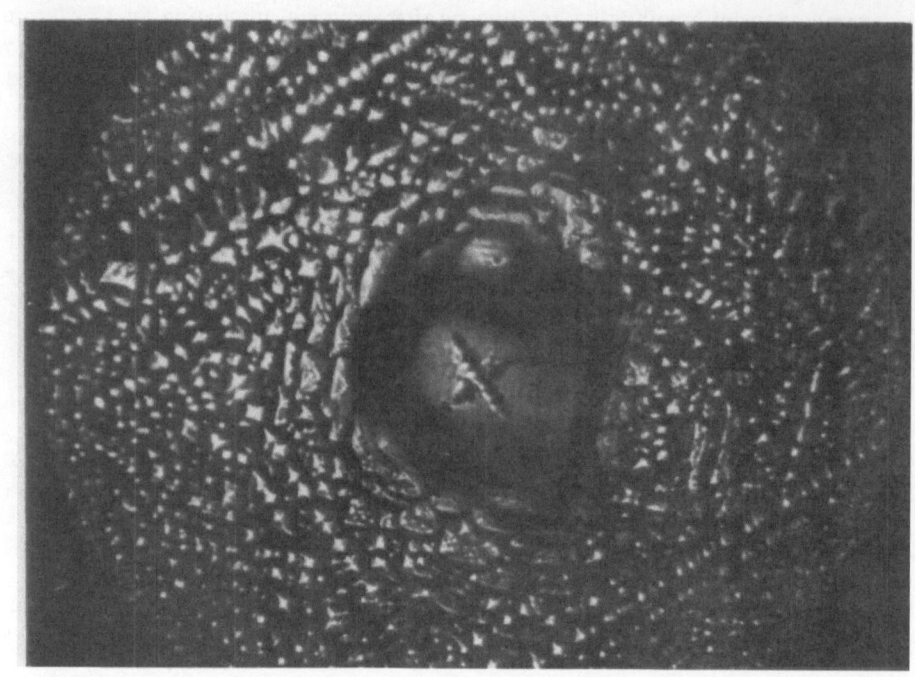

Fig. 62. Chains of etch figures at the bottom of a pit, × 270.

Fig. 61. Etch pits on a crystal of LiF after prolonged insonation, × 585.

pits occur at the bottom of the large one; the directions of these coincide with the directions of chains on the faces (Fig. 62).

Plastically deformed crystals also give etch figures; for example, a scratch may be made with a needle before the LiF crystal is irradiated, which causes the pits to have a higher density along this line (Fig. 63). Again, a crystal exposed in the spark gap of an induction coil shows more etch pits near the points where sparks have passed.

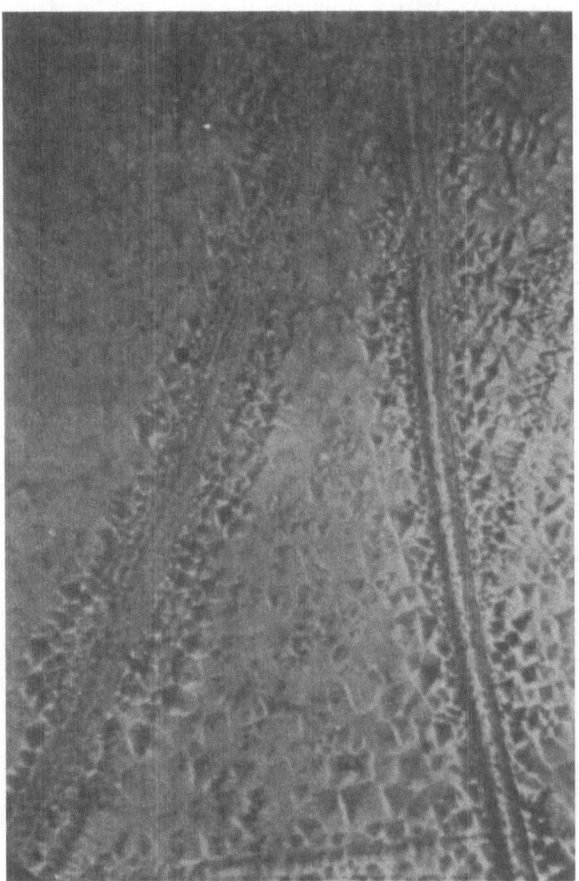

Fig. 63. Etch figures on the surface of a plastically deformed crystal of LiF, × 270.

Ultrasonic etching produces similar effects with crystals of terpene monohydrate, potassium dichromate, TlI + TlBr, and so on.

If the etch pits are correlated with the points of emergence of dislocations, then the patterns can be inter-preted as implying that dislocations are present throughout the volume of the crystal.

8. Effects of Ultrasound on Liquid Crystals

The following are some experimental results on the behavior of liquid crystals in ultrasonic fields.

Liquid crystals are so called because they combine the properties of liquid and crystal. On heating, they melt at a definite temperature to give a normal isotropic liquid; they are anisotropic (birefringent) and some-times differ in shape from ordinary drops of liquid. Further, they give regular overgrowths with true crystals and may be oriented by electric and magnetic fields.

On the other hand, they can flow and do not have lattice structures and so on [4].

Liquid crystals are organic or biological systems; we now know of about 3000 substances that can exist in the liquid-crystal state.

The molecules of a substance that can give a liquid crystal are usually elongated; the structure is governed by the mutual positions of the molecules in the liquid crystal. Structures resembling those of liquid crystals occur in some polymers and colloidal solutions.

Friedel [83] has shown that an important feature of a liquid crystal is that the detailed structure is intermediate between the three-dimensional lattice structure of a solid and the structure of an amorphous liquid. There are two types of liquid crystal; in the first, the molecules are parallel and have short-range order (nematic crystals), while in the second they are arranged in layers, each layer consisting of parallel molecules showing short-range order (smectic crystals).

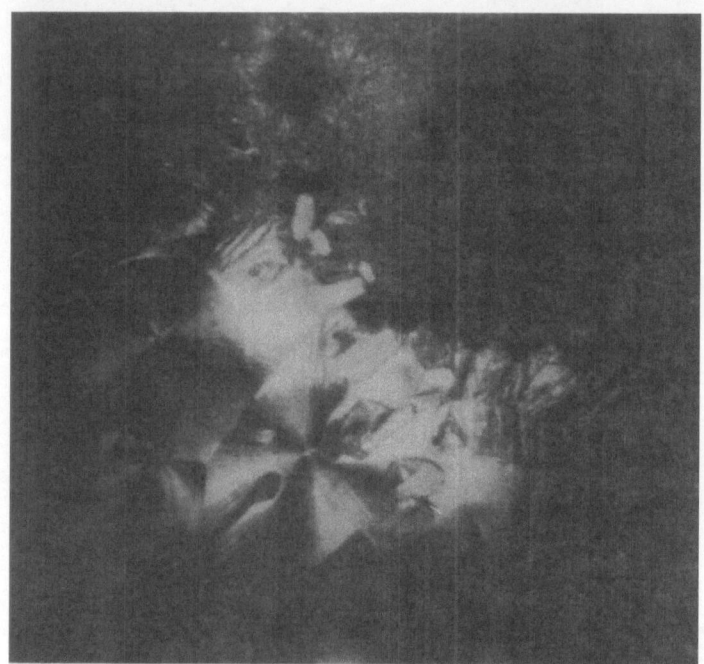

Fig. 64. Structure produced in cholesteryl acetate by ultrasound.

There is considerable interest in liquid crystals at present, for they throw much light on the structure of matter and provide a connection between the physics of liquids and the properties of solids. Many soaps and detergents can exist in liquid-crystal form, and the theory of liquid crystals enables us to make the best use of their properties. It is to be expected that liquid crystals will be found to show piezoelectric and pyroelectric properties.

Shubnikov has stated that "it is necessary to establish how to make homogeneous mesomorphic bodies in the supercooled (vitreous) state in large volumes. These mesomorphic monocrystals should be of practical value in utilizing the unusual properties of liquid crystals such as the enormous optical activity (many hundreds of complete turns in 1 mm), the strong pleochroism, and the pronounced electrical and magnetic properties."

We have used cholesteryl acetate, ethyl p-azoxybenzoate, ethyl p-azoxycumarate, p-azoxyphenetol, and p-azoxyanisole.

A small quantity of the material was placed on a slide under a coverslip and was slowly heated to its melting point on a hotplate. The preparation was observed with the polarizers parallel or crossed. Transitions from the isotropic state to the liquid-crystal state and further to the solid state can be observed. The melt was allowed to cool naturally, so no temperature measurements were made. In some cases, fixed temperatures were maintained within 1-2° by a regulator.

Fig. 65. Ethyl p-azoxybenzoate crystallized in an ultrasonic field.

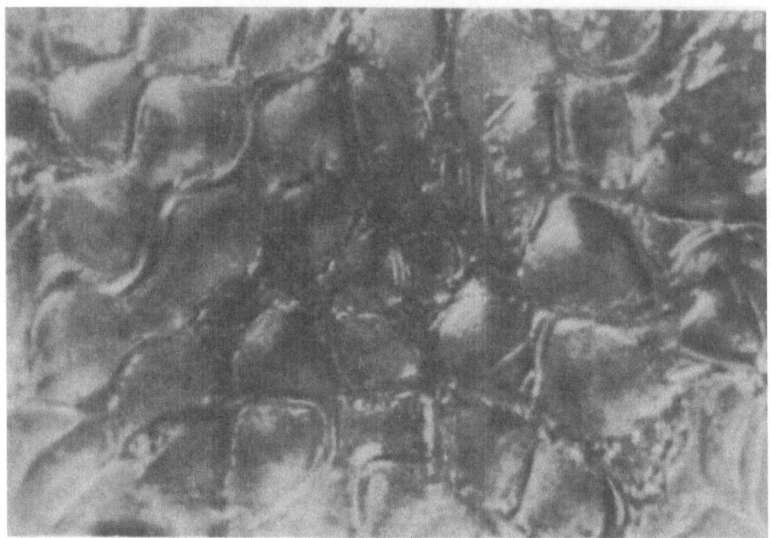

Fig. 66. Cellular structure of p-azoxyphenetol produced by ultrasound
in a thin film.

The liquid crystals were exposed to ultrasonic fields, which were applied through a 30 × 30 mm glass waveguide 60 mm long standing directly on the quartz crystal (720 kc). The crystal and guide were mounted on the microscope stage.

Cholesteryl acetate was prepared as above and was set up on the waveguide; good acoustic contact was provided by a thin film of oil. A small electric heater at the top of the waveguide provided the necessary temperature.

The ultrasound was turned on when the preparation was entirely in the liquid-crystal state; the field caused a peculiar circulation in the preparation, and the final solid structure always differed greatly from that produced in the absence of the field. The structure was fine-grained where the field was strongest, the usual spherulite structure occurring at points of lower intensity (Fig. 64).

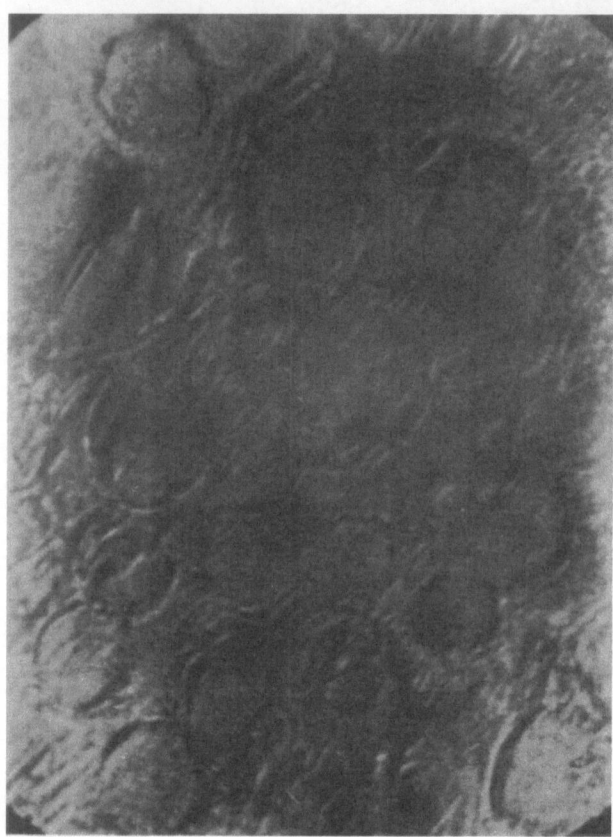

Fig. 67. Cellular structure of p-azoxyphenetol preserved in the
solid state.

Ethyl p-azoxybenzoate gives a smectic phase; here the field was applied when the solid crystals had begun to form. Much of the liquid-crystal part then began to move vigorously. The final solid consisted in part of needles, the spaces between these being filled by small crystals all of much the same size (Fig. 65). The color of the needles as seen in polarized light varied as the stage was rotated; the small crystals did not show interference colors. If the layer of these small crystals was very thin, it was possible to see under them other crystals that did show interference colors. High intensities at 720 kc produced dispersal of the solid crystals growing in the liquid-crystal phase. Some interesting effects occurred in p-azoxyphenetol as it cooled from the isotropic state. The material was melted on the slide and then was enclosed under a coverslip; the layer was made uniform by remelting the material. The nematic phase gave the appearance of filaments under the microscope.

Zocher [84, 85] considers that these filaments represent points of optical discontinuity (points where the optic axes of the molecules change direction sharply). The filaments are firmly attached to the glass in places. The ultrasound (720 kc) broke the filaments, which began to move about; a cellular structure appeared in a few seconds (Fig. 66). The layer was split up into cells, about which flowed the liquid-crystal phase. These microscopic vortices persisted down to the temperature at which the solid appeared. The cellular structure persisted for a while after the field was turned off; it returned in its previous form when the field was turned on again. Sometimes it persisted even in the solid state (Fig. 67).

A detailed study of liquid crystals in this way should give rise to many new results of theoretical and practical interest.

Literature Cited

1. L. Bergmann, Ultrasonics [Russian translation](Moscow, Izd-vo inostr. lit., 1956).
2. W. Cady, Piezoelectricity: an Introduction to the Theory and Applications of Electromechanical Phenomena in Crystals [Russian translation] (Moscow, Izd-vo inostr. lit., 1949).
3. N. A. Roi, "Origin and Development of Ultrasonic Cavitation." Akust. zh., $\underline{3}$, 3 (1957).
4. A. V. Shubnikov, How Crystals Grow [in Russian] (Moscow-Leningrad, Izd-vo AN SSSR, 1935).
5. A. V. Shubnikov, Crystals in Science and Technology [in Russian] (Moscow, Izd-vo AN SSSR, 1958).
6. H. Yasuchi, Nippon kagaku deassi, $\underline{73}$, 114 (1952).
7. V. D. Kuznetsov, Crystals and Crystallization [in Russian] (Moscow, Gostekhizdat, 1953).
8. E. Hiedemann, "Metallurgical Effects of Ultrasonic Waves." Acoust. Soc. Am., $\underline{26}$, 5 (1954).
9. I. G. Mikhailov, Propagation of Ultrasonic Waves in Liquids [in Russian] (Gostekhizdat, 1949).
10. B. B. Kudryavtsev, Use of Ultrasonic Methods in Physicochemical Research [in Russian] (Moscow-Leningrad, Gostekhizdat, 1952).
11. V. F. Nazdrev, Use of Ultrasonics in Molecular Physics [in Russian] (Moscow, Gostekhizdat, 1958).
12. V. I. Danilov, Problems of Metal Science and the Physics of Metals [in Russian] (Moscow, Metallurgizdat, 1949).
13. A. Crawford, Ultrasonic Engineering [Russian translation] (Moscow, Izd-vo inostr. lit., 1958).
14. L. G. Merkulov, "Calculation of Ultrasonic Concentrators." Akust. zh., 3, No. 3, 230-238 (1953); "Theory and Calculation of Composite Concentrators." Akust. zh., $\underline{5}$, No. 2, 183-190 (1959).
15. K. A. Nagol'nykh and L. D. Rozenberg, "The Optimal Mode of Operation of a Power Concentrator." Akust. zh., $\underline{6}$, 3 (1960).
16. W. T. Richards and A. L. Loomis, "The Chemical Effects of High-Frequency Sound Waves." J. Am. Chem. Soc., $\underline{49}$, 3086 (1927).
17. R. W. Wood and A. L. Loomis, "The Physical and Biological Effects of High-Frequency Sound Waves of Great Intensity." Phil. Mag., $\underline{4}$, 417 (1927).
18. V. I. Danilov, E. Pluzhnik, and B. Teverovskii, "Generation of Crystallization Centers in a Supercooled Liquid." Zh. éksperim. i teor. fiz., $\underline{9}$, 66 (1939).
19. V. I. Danilov and B. Teverovskii, "Generation of Crystallization Centers in a Supercooled Liquid." Zh. éksperim. i teor. fiz., $\underline{10}$, 11 (1940).
20. R. Ya. Berlaga, "Effects of an Ultrasonic Field on the Crystallization of Supercooled Liquids." Zh. éksperim. i teor. fiz., $\underline{9}$, 11 (1939).
21. R. Ya. Berlaga, "Effects of an Ultrasonic Field on the Crystallization of Supercooled Liquids." Zh. éksperim. i teor. fiz., $\underline{16}$, 7 (1946).
22. G. L. Mikhnevich and P. N. Dombrovskii, "Effects of Low-Frequency Elastic Vibrations on the Crystallization of a Supercooled Organic Liquid." Zh. éksperim. i teor. fiz., $\underline{10}$, 2 (1940).
23. S. V. Belynskii, "Crystallization of Salts." Zh. éksperim. i teor. fiz., $\underline{6}$, 10 (1936).
24. P. N. Shablykin, "Crystallization of Sulfur in the Field of Ultrasonic Waves." Mineral'noye syr'e, $\underline{12}$, (1937).
25. M. M. Mazhul', "Cavitation Phenomena and the Formation of Crystal Nuclei." (Dissertation, 1954).
26. S. Ya. Sokolov, Sur L'influence des ondes ultra-soniques sur les reactions chimiques. Techn. Phys. URSS, $\underline{3}$, 176 (1936).
27. I. T. Sokolov, "Effects of Ultrasonics on Supercooled Water." Zh. tekh. fiz., $\underline{8}$, 10, 901 (1938).

28. V. I. Danilov and G. Kh. Chedzhemov, Problemy metallovedeniya i fiziki metallov, 4 (1955).

29. P. Günther and W. Zeil, "Die Kristallisationsgeschwindigkeit von Glycerin und von Benzophenon im Ultraschallfeld." Z. anorg. allg. Chem., 285, No. 3/6, 191 (1956).

30. C. Turner, T. Galkowski, W. Radle, and A. Wanhook, "Grain Formation by Sonic Irradiation." The International Sugar Journal, 22, 621, 298 (1950).

31. S. Ya. Sokolov, L'influence des ondes ultra-acoustiques sur le procés de solidification des métaux fondus. Acta Physicochim. URSS, 3, 939 (1935).

32. G. Schmid and L. Ehret, "Die Wirkung intensiven Schalls auf Metallschmelzen." Z. Elektrochem., 43, 869 (1937).

33. G. Schmid and A. Roll, "Die Wirkung intensiven Schalls auf Metallschmelzen." Z. Elektrochem., 45, 769 (1939).

34. G. Schmid and A. Roll, "Die Wirkung intensiven Schalls auf Metallschmelzen." Z. Elektrochem., 46, 653 (1940).

35. H. Seemann, "Metallforschung mit Ultraschall." Metallwirtsch., 15, 1067 (1936).

36. O. Nomoto, "Die Wirkung ultraakustischer Schwingungen auf die Metallschmelzen." J. Japan Inst. Metals, 8, 54 (1944).

37. A. P. Kapustin, "Effects of Ultrasonics on the Structure of Some Nonferrous Metals." Uchenye zapiski MGPI, 1 (1947); "An Experimental Study of the Effects of Ultrasonics on the Kinetics of Crystallization." Dissertation (1951).

38. Ya. B. Gurevich, V. I. Leont'ev, and I. I. Teumin, "Effects of Ultrasonics on the Structure and Properties of a Steel Casting." Stal', 5, 406 (1957).

39. I. I. Teumin, All-Union Conference on the Use of Ultrasonic Techniques in Industry [in Russian] (1957).

40. N. P. Nikolaichik and E. N. Nikolaichik, All-Union Conference on the Use of Ultrasonic Techniques in Industry [in Russian] (1957).

41. N. P. Nikolaichik and E. N. Nikolaichik, "Improvement of Castings by Means of Ultrasonics." Stal', 4, 322 (1957).

42. W. Rostoker and M. Berger, Foundry, 81, 100, 260 (1953).

43. H. Seemann and H. Menzel, Zs. Metall., 1, 39 (1947).

44. I. G. Polotskii and Ya. Benieva, All-Union Conference on the Use of Ultrasonic Techniques in Industry [in Russian] (1957).

45. G. Siebers and W. Bulian, Metallforschung, 1, 158 (1946).

46. H. Uedzawa, "A Study of the Effects of Ultrasonic Waves on the Structure of Al — Si Alloys." J. Japan. Inst. Metals, 23, 3, 168 (1959).

47. N. N. Sirota, E. A. Lekhtblau, and E. M. Smolyarenko, "The Effects of Ultrasonic Energy Applied during Crystallization on the Structure and Properties of Alloys of Aluminum with Silicon." Fiz. metal. i metalloved., 7, 6, 879 (1959).

48. M. V. Klassen-Neklyudova and A. P. Kapustin, "Effects of Ultrasonic Energy on the Stress Distribution in a Monocrystal of a Solid Solution of Thallium Bromide and Iodide." Dokl. AN SSSR, 27, No. 6 (1951).

49. A. P. Kapustin, "Effects of Ultrasonic Energy on the Polymorphic Transformation of Ammonium Nitrate." Dokl. AN SSSR, 26, No. 3 (1951).

50. H. Hollmann and W. Bauch, "Der magnetische Barkhausen-Effekt bei Ultraschallbestrahlung." Naturwiss., 23, 35 (1935).

51. G. Schmid and U. Jetter, "Einfluss von Ultraschall auf das magnetische Verhalten von Nickel." Zs. Electrochem., 47, 155 (1941).

52. M. Mahoux, Mechanique, 21, 281 (1937).

53. A. P. Kapustin, Use of Ultrasonics in the Examination of Materials. Action of Ultrasonics on Plexiglas, Issue 6 [in Russian] (Moscow, Izd-vo MOPI, 1958).

54. L. D. Rozenberg, Applications of Ultrasonics [in Russian] (Izd-vo AN SSSR, 1957).

55. A. P. Kapustin, "Effects of Ultrasonic Energy on Rates of Phase Transition in Organic Substances." Zh. tekh. fiz., 20, 10, 1158 (1950); "Crystallization of Organic Substances under the Influence of Ultrasound." Zh. tekh. fiz., 22, 5, 765 (1952); "An Experimental Study of the Effects of Ultrasonic Energy on the Kinetics of Crystallization." Izv. AN SSSR, 14, 3, 357 (1950).

56. S. N. Rzhevkin and E. P. Ostrovskii, "Emulsification by Means of Ultrasonics." Zh. tekh. fiz., 6, 73 (1935).

57. Kh. S. Bagdasarov, "An Experimental Study of Processes of Crystallization and Dissolution in an Ultrasonic Field." Dissertation (1957).

58. A. Shubnikov and G. Lemmlein, "Beobachtungen über die Orthotropie des Kristallwachstums." Z. Krist., 65, 297 (1927).

59. A. A. Bochvar, A Study of the Mechanism and Kinetics of the Crystallization of Alloys of Eutectic Type [in Russian] (Moscow-Leningrad, ONTI, 1935).

60. A. P. Kapustin, "Production of Textures in Crystalline Substances by Means of Ultrasonic Waves." Dokl. AN SSSR, 1, No. 3, 21 (1950).

61. I. G. Polotskii, Dissertation, Institute of Ferrous Metallurgy, Academy of Sciences of the Ukr. SSR (1942).

62. A. P. Kapustin and V. E. Kavalyunaite, Use of Ultrasonics in the Examination of Materials. Effects of Vibration of the Walls of the Vessel on Crystallization in Thin Layers [in Russian] (Izd-vo MOPI, issue 9, 1959).

63. Ya. I. Frenkel', The Kinetic Theory of Liquids [in Russian] (Izd-vo AN SSSR, 1945).

64. A. P. Kapustin, "Spontaneous Production of Crystallization Centers in an Ultrasonic Field." Uchenye zapiski MGPI im. Lenina, 38 (1954).

65. G. Tamman, Aggregatzustände. Leipzig (1922).

66. A. P. Kapustin, "Effects of Ultrasonic Energy on Nucleation in Vitreous Materials." Uchenye zapiski MGPI im. Lenina, 3, 51 (1957).

67. A. P. Kapustin and V. E. Kavalyunaite, "Effects of Ultrasonic Energy on the Growth of Potash Alum Mono-crystals." Kristallografiya, 1, 6, 737 (1956); "Crystallization of Alum from Aqueous Solution in an Ultra-sonic Field." Growth of Crystals, Vol. 2, Moscow, Izd-vo AN SSSR (1957) [English translation: New York, Consultants Bureau, 1960].

68. M. P. Shaskol'skaya and A. V. Shubnikov, "Uber die künstliche Herstellung gesetzmässiger Kristallver-wachsungen des Kalialauns." Zeitschr. f. Krist., 85, 1 (1933).

69. Kh. S. Bagdasarov, G. V. Berezhkova, and A. P. Kapustin (in press).

70. G. Pfefferkorn, "Elektronenmikroskopische Untersuchungen an Kalkspat und dessen Realkristallbau." Optik, 7, 208 (1950).

71. A. P. Kapustin and M. A. Fomina, "Effects of Ultrasonic Energy on the Dissolution of Steel in Sulfuric Acid." Dokl. AN SSSR, 1, No. 6, 33, 847 (1952).

72. I. V. Tsyganova and V. M. Lebedev (in press).

73. S. Yamaguchi, "Metallic Corrosion Influenced by Ultrasonic Waves." J. Appl. Phys., 23, 1057 (1952).

74. W. Meiswinkel, V.D.I., 97, 2, 42 (1955).

75. A. Honess, The Nature, Origin, and Interpretation of the Etch Figures of Crystals. New York (1927).

76. Kh. S. Bagdasarov and A. P. Kapustin, "Production of Etch Figures with Ultrasonic Vibrations." Kristallo-grafiya, 1, No. 1, 139 (1956).

77. Kh. S. Bagdasarov and V. Ya. Khaimov-Mal'kov, "Some Experimental Data on the Cause of the Formation of Etch Figures in an Ultrasonic Field." Kristallografiya, 2, 2 (1957).

78. N. P. Kameneva, Use of Ultrasonics in the Examination of Materials, Issue 6 [in Russian] (Moscow, Izd-vo MOPI, 1958).

79. Kh. S. Bagdasarov, "The Effects of Ultrasonic Vibrations on the Dissolution of Monocrystals." Kristallo-grafiya, 3, No. 1 (1958).

80. J. J. Gilman and W. Johnston, "Observations of Dislocation Glide and Climb in Lithium Fluoride Crystals." J. Appl. Phys., 27, 1018 (1956).

81. A. P. Kapustin, "Detection of Dislocations by Means of Ultrasonics." Kristallografiya, 4, No. 2, 265 (1959); "Etch Figures in Crystals of Terpene Monohydrate." Use of Ultrasonics in the Examination of Materials [in Russian] (Moscow, Izd-vo MOPI, Issue 9, 1959).

82. N. N. Sheftal', Growth of Crystals, Vol. 1, Izd-vo AN SSSR (1957) [English translation: New York, Consultants Bureau, 1958].

83. G. Friedel and E. Friedel, Les propriétés physiques des stases mésomorphes général et leur importance com-me principe de classification, Z. Krist., 79, 1/4 (1931).

84. H. Zocher and V. Birstein, Zeitschr. Physik Chem., A., 142, 186 (1929).

85. H. Zocher, "Über die Kontinuumtheorie und die Schwarmtheorie der nematischen Phasen." Ann. Physik, 31, 570 (1938).